Nuclear Physics: A Very Short Introduction

VERY SHORT INTRODUCTIONS are for anyone wanting a stimulating and accessible way into a new subject. They are written by experts, and have been translated into more than 40 different languages.

The series began in 1995, and now covers a wide variety of topics in every discipline. The VSI library now contains over 400 volumes—a Very Short Introduction to everything from Psychology and Philosophy of Science to American History and Relativity—and continues to grow in every subject area.

Very Short Introductions available now:

ACCOUNTING Christopher Nobes
ADVERTISING Winston Fletcher
AFRICAN AMERICAN RELIGION
    Eddie S. Glaude Jr.
AFRICAN HISTORY John Parker and
    Richard Rathbone
AFRICAN RELIGIONS Jacob K. Olupona
AGNOSTICISM Robin Le Poidevin
ALEXANDER THE GREAT
    Hugh Bowden
AMERICAN HISTORY Paul S. Boyer
AMERICAN IMMIGRATION
    David A. Gerber
AMERICAN LEGAL HISTORY
    G. Edward White
AMERICAN POLITICAL HISTORY
    Donald Critchlow
AMERICAN POLITICAL PARTIES
    AND ELECTIONS L. Sandy Maisel
AMERICAN POLITICS Richard M. Valelly
THE AMERICAN PRESIDENCY
    Charles O. Jones
THE AMERICAN
    REVOLUTION Robert J. Allison
AMERICAN SLAVERY
    Heather Andrea Williams
THE AMERICAN WEST Stephen Aron
AMERICAN WOMEN'S HISTORY
    Susan Ware
ANAESTHESIA Aidan O'Donnell
ANARCHISM Colin Ward
ANCIENT ASSYRIA Karen Radner
ANCIENT EGYPT Ian Shaw
ANCIENT EGYPTIAN ART AND
    ARCHITECTURE Christina Riggs

ANCIENT GREECE Paul Cartledge
THE ANCIENT NEAR EAST
    Amanda H. Podany
ANCIENT PHILOSOPHY Julia Annas
ANCIENT WARFARE Harry Sidebottom
ANGELS David Albert Jones
ANGLICANISM Mark Chapman
THE ANGLO-SAXON AGE John Blair
THE ANIMAL KINGDOM
    Peter Holland
ANIMAL RIGHTS David DeGrazia
THE ANTARCTIC Klaus Dodds
ANTISEMITISM Steven Beller
ANXIETY Daniel Freeman and
    Jason Freeman
THE APOCRYPHAL GOSPELS
    Paul Foster
ARCHAEOLOGY Paul Bahn
ARCHITECTURE Andrew Ballantyne
ARISTOCRACY William Doyle
ARISTOTLE Jonathan Barnes
ART HISTORY Dana Arnold
ART THEORY Cynthia Freeland
ASTROBIOLOGY David C. Catling
ATHEISM Julian Baggini
AUGUSTINE Henry Chadwick
AUSTRALIA Kenneth Morgan
AUTISM Uta Frith
THE AVANT GARDE David Cottington
THE AZTECS David Carrasco
BACTERIA Sebastian G. B. Amyes
BARTHES Jonathan Culler
THE BEATS David Sterritt
BEAUTY Roger Scruton
BESTSELLERS John Sutherland

Available soon:

For more information visit our website

www.oup.com/vsi/

Frank Close

NUCLEAR PHYSICS

A Very Short Introduction

OXFORD
UNIVERSITY PRESS

OXFORD
UNIVERSITY PRESS

Great Clarendon Street, Oxford, OX2 6DP,
United Kingdom

Oxford University Press is a department of the University of Oxford.
It furthers the University's objective of excellence in research, scholarship,
and education by publishing worldwide. Oxford is a registered trade mark of
Oxford University Press in the UK and in certain other countries

Published in the United States of America by Oxford University Press
198 Madison Avenue, New York, NY 10016, United States of America

British Library Cataloguing in Publication Data
Data available

Library of Congress Control Number: 2015931639

ISBN 978-0-19-871863-5

Printed in Great Britain by
Ashford Colour Press Ltd, Gosport, Hampshire

Links to third party websites are provided by Oxford in good faith and
for information only. Oxford disclaims any responsibility for the materials
contained in any third party website referenced in this work.

Contents

List of illustrations

Chapter 1
The fly in the cathedral

Atoms

When we look at the world around us, there is nothing that immediately shows that everything is made of atoms, let alone that at the heart of every atom is a compact nucleus, where powerful forces are at work.

In 1808, the English chemist John Dalton proposed the modern atomic theory of matter.

Chemistry was becoming a quantitative science and had shown that a wide variety of substances could be formed by combining different quantities of a few elements such as hydrogen, carbon, and oxygen. Dalton realized that if each element is made from atoms—very tiny objects, which join together to build up the substances that are large enough to see—he could explain these regularities. Combining atoms of various elements together made molecules of non-elementary substances. Furthermore, he believed that atoms were indivisible; indeed, it was for this reason that he named them 'atoms', in honour of the philosophers of ancient Greece who coined 'atomos'—indivisible.

Ninety-eight of these elementary substances are now known to occur naturally on earth, some more familiar than others.

(The origin of 98 is explained in chapter 5). With every breath, you inhale a million billion billion atoms of oxygen, which gives you some idea of how small each one is. Atoms are not the smallest things, however. Whereas little more than a century ago they were thought to be small impenetrable objects, like miniature billiard balls perhaps, today we know that each has a rich labyrinth of inner structure, where electrons whirl round a massive compact central nucleus. It is the gyration of the electrons, switching ephemerally between neighbouring atoms, that links one to another, building up molecules and bulk matter.

Electrons are held in place, remote from the nucleus, by the electrical attraction of opposite charges, electrons being negatively and the atomic nucleus positively charged. A temperature of a few thousand degrees is sufficient to break this attraction completely and liberate all of the electrons from within atoms. Even room temperature can be enough to release one or two; the ease with which electrons can be moved from one atom to another is the source of chemistry, biology, and life. Restricted to conditions that are cool relative to those of the stars and modern physics, the 19th-century scientist was only aware of chemical activity; the heart of the atom—the nucleus—was hidden from view.

The first indirect hints that atoms are not fundamental emerged around 1869, when Dmitri Mendeleev discovered that when he made a list of the atomic elements from the lightest—hydrogen—up to the heaviest then known—uranium—elements with similar properties would recur at regular intervals (Figure 1). If each element were truly independent then any similarities among them would have occurred at random. The empirical periodicity was striking. Today we know the cause: atoms are not elementary, but are instead complex systems built from common constituents; electrons surround a compact nucleus. The experiments that led to this picture of atomic structure were made by Ernest Rutherford and his co-workers little more than 100 years ago and were the fruits of a sequence of discoveries spanning several years.

1 H																		2 He
3 Li	4 Be												5 B	6 C	7 N	8 O	9 F	10 Ne
11 Na	12 Mg												13 Al	14 Si	15 P	16 S	17 Cl	18 Ar
19 K	20 Ca		21 Sc	22 Ti	23 V	24 Cr	25 Mn	26 Fe	27 Co	28 Ni	29 Cu	30 Zn	31 Ga	32 Ge	33 As	34 Se	35 Br	36 Kr
37 Rb	38 Sr		39 Y	40 Zr	41 Nb	42 Mo	43 Tc	44 Ru	45 Rh	46 Pd	47 Ag	48 Cd	49 In	50 Sn	51 Sb	52 Te	53 I	54 Xe
55 Cs	56 Ba	57–70 *	71 Lu	72 Hf	73 Ta	74 W	75 Re	76 Os	77 Ir	78 Pt	79 Au	80 Hg	81 Ti	82 Pb	83 Bi	84 Po	85 At	86 Rn
87 Fr	88 Ra	89– * *																

* Lanthanide series	57 La	58 Ce	59 Pr	60 Nd	61 Pm	62 Sm	63 Eu	64 Gd	65 Tb	66 Dy	67 Ho	68 Er	69 Tm	70 Yb
** Actinide series	89 Ac	90 Th	91 Pa	92 U										

1. The periodic table of chemical elements.

Nature has buried its secrets deep, but not entirely hidden them. Clues to the restless agitation within its atomic architecture are everywhere. The static electricity released when brushing dry hair, the effects of lightning, the aurora borealis caused by cosmic particles hurtling towards the north magnetic pole, and the ability of a small compass needle to sense magnetism that is thousands of miles away are just a few examples of the electrical forces within atoms. The radioactivity of natural rocks and the ability of radium to pour out energy and remain warmer than its surroundings are the results of powerful forces at work deeper still: within the atomic nucleus.

Wilhelm Roentgen's discovery of X-rays in 1895 marks the start of modern atomic and nuclear physics. Today we know that X-rays are a form of electromagnetic radiation—essentially light whose wavelengths are much shorter than those that our eyes record as colours—but at the end of the 19th century the challenge was to explain what X-rays were and how they were formed. In the course of sorting this out a host of other discoveries were made, most notably that atoms have a complex internal structure.

In Paris, Henri Becquerel learned of Roentgen's X-rays in the first week of January 1896. He wondered if X-rays and phosphorescence were related. To find out he planned to activate some phosphorescent crystals by exposure to sunlight for several hours. First he wrapped them in opaque paper, placed the package on top of some photographic emulsion, and then put them all in a dark drawer. If the crystals emitted only visible light, none would get through the opaque paper to reach the photographic emulsion, whereas any X-rays would pass uninterrupted to the plates and fog them. As an extra test he placed some metal pieces between the package and the photographic material so that even X-rays would be blocked and leave a silhouette of the metal in the resulting image.

Becquerel was doubly fortunate. First, his crystals contained uranium; second, Paris was overcast for weeks, which prevented

4

him from putting his plan into action. By 1 March he was becoming frustrated. At this point, for want of something to do, he decided to develop the plates anyway, expecting, as he later reported, to find a feeble image at best.

To his surprise the images were sharp, which showed that uranium radiates energy spontaneously without need of prior stimulation such as sunlight.

Today, this discovery of spontaneous 'radioactivity' is recognized as seminal, but at the time it made no special impact. The real birth of the radioactive era was when Pierre and Marie Curie discovered radioactivity in other elements, in particular radium. Here the effect is so powerful that samples of the element glow in the dark. (It was Marie Curie who invented the term *radioactivity*.) By 1903 the significance was fully realized and it was appropriate that the Curies shared the Nobel Prize with Becquerel. It would be half a century before the source of radioactivity was fully understood, but in the interim the phenomenon had been put to use by Ernest Rutherford, and others, as a tool to reveal atomic structure.

An α, β, γ of radioactivity

The end of the 19th century was a glorious time for atomic physics. Following Roentgen's discovery of X-rays, and Becquerel's radioactivity, in 1897 J.J. Thomson discovered the electron. He realized that electrons are common to all atoms, which implied that every atom has an internal structure.

Electrons carry electric charge, deemed *negative* by convention, and this immediately suggested a means for atoms to be built: the attraction of opposite charges provides the force that can bind negatively charged electrons to some positively charged entities. The question was: what carries the positive charge and how is it distributed within the atom?

5

Thomson thought that the mass of an atom is due to its electrons, which meant that several thousand electrons were required to make an atom of hydrogen, the lightest example. In his 'plum pudding' model, the electrons were distributed like plums in some amalgam of positive charges—the pudding.

In October 1895, one month before Roentgen discovered X-rays and two months before the dramatic announcement of their discovery, a young New Zealander, Ernest Rutherford, had left home and travelled halfway around the world to England and the Cavendish laboratory in Cambridge. Under Thomson's direction, Rutherford began to investigate how the newly discovered X-rays ionize gases. In the course of this, Rutherford established that there is more than one form of radioactivity.

He had begun by studying the ionization produced by the radiation from uranium, but then changed tack: he decided to use the ionization to learn about radioactivity. To do so he used an electrometer, whose basic idea is to measure the deflection of a charged metallic strip in an electric field.

Rutherford investigated how the radioactivity is absorbed by various materials. He covered the uranium with sheets of aluminium and discovered that the radiation is progressively absorbed. About one hundredth of a millimetre of foil was enough to reduce the radiation considerably, but after this, as he added more foil the radiation appeared to maintain its strength. Only after he had added several millimetres of aluminium did the radiation's intensity markedly die away again.

He deduced from this that there are two components to the radiation: one, which he called *alpha*, is easily absorbed; the other, which penetrated the aluminium foil more easily, he named *beta*.

Subsequently, a third variety was found, distinct from the previous two. This became known as *gamma* radiation. Today we know that gamma rays are a form of high-energy X-ray. While X-rays may be emitted by electrons bound deep in large atoms, gamma ray are emitted from the atomic nucleus. Rutherford would later show that the beta rays consist of electrons—not ones that pre-exist within an atom but ones that are formed when atomic nuclei adjust their stability (chapter 2). He also later showed that alpha particles are the nuclei of helium atoms.

In the process of identifying the two forms of radiation, in 1900 Rutherford had noticed something peculiar: his radioactive source—thorium—was emitting a gas that was itself radioactive. This appeared to be the same gas that Pierre and Marie Curie had discovered being emitted by radium in 1898. In identifying this gas, Rutherford and his collaborator, Frederick Soddy, discovered atomic alchemy, for which they won the Nobel Prize.

Alchemy

By this stage, 1900, Rutherford was based at McGill University in Montreal. Frederick Soddy, newly arrived from Oxford, was a chemist. He analysed the gas, and after a series of detailed investigations, which involved the separation of different radioactive materials, established that the emanation was not only a gas but a new element, chemically inert like argon, now known as radon. This was revolutionary. Two elements, radium and thorium, hitherto believed to be immutable fundamental foundations of matter—atoms indeed—could spontaneously emit atoms of radon, another supposedly fundamental element. More surprises were in store. Soddy found that the thorium transmutes first to a form of radium, and finally to radon, at each stage spitting out radiation.

Not only had they established that the decay products consisted of a variety of elements, but also that these were physically variant forms of the familiar elements. These different forms of an element are chemically identical—hence the same element—but physically different. These distinct varieties are called isotopes, from the Greek for 'equally placed' (in the periodic table of the elements).

The chemistry of different isotopes is the same, but their physical properties can vary widely. One isotope of a given element might be stable, whereas another is radioactive, and decays. In general, isotopes of a single element have a range of half-lives—the term *half-life* being the time it takes for half of a radioactive sample to decay. It was the vast range of half-lives of the products in the thorium decays that enabled Soddy to identify their distinct identities. The origin of isotopes will be described in chapter 2.

Around this time, Pierre and Marie Curie had discovered new radioactive elements—radium and polonium—in the products of uranium's disintegration. Their breakthrough, together with the work of Rutherford and Soddy, were the clues that atoms have an inner structure which differs only slightly between one atom and the next. Small changes in this inner structure would convert one type of atom into another.

A first step to solving this puzzle came when Rutherford established the nature of the alpha particle.

In 1905, he studied the alpha radiation produced by the decays of thorium, radium, and other elements. He found that the alpha particles in each case had the same mass as an atom of helium, but unlike normal helium atoms, which are electrically neutral, an alpha particle carries two units of positive charge (in the convention where an electron is negatively charged). He suspected, correctly, that an alpha particle is a doubly ionized atom of helium. This also explains the mystery of why helium is found trapped within crystalline ores of uranium and thorium: it is because those radioactive elements are spontaneously emitting alpha particles.

Having established that alpha particles have positive charge, and a mass similar to that of light atoms, Rutherford had an epiphany:

by bombarding atoms with alpha particles, and seeing how the particles scatter, he could learn about atomic structure.

Electrons, alphas, and nuclei

J.J. Thomson had shown that electrons are almost two thousand times lighter than the lightest atom (hydrogen). If electrons were one of the elementary ingredients that built up atoms, what else is there? How are the positive and negative charges arranged within atoms? What causes transmutation? What is the source of radioactivity? The challenge at the start of the 20th century was to unravel the atomic structure.

First it was necessary to identify precisely what carried the positive charge in atoms, and this needed some way of looking inside them. Rutherford's work with alpha particles gave him the means, and with brilliant directness, he went straight to the heart of the problem.

Alpha particles are emitted from atoms and so are very small. They move at about 15,000 kilometres per second, which is about one-twentieth of the speed of light, yet Rutherford noticed that a thin sheet of mica could deflect them slightly. He calculated that the electric forces within the mica must be immensely powerful compared with anything then known, and deduced that these powerful electric fields must exist only within exceedingly small regions, smaller even than an atom. From this came his inspired guess: these intense electric fields are what hold the electrons in their atomic prisons and are capable of deflecting the swift alpha particles. From this came his big idea: from the way that atoms deflect the beams of speeding alpha particles, he would be able to deduce the atom's electrical structure.

Alpha particles themselves have positive charge, but Rutherford knew that they could not be the source of the positive charge of all atoms: their mass is about four times that of hydrogen and their charge is twice that of the electron (and of opposite sign, of course), so you couldn't build hydrogen with an alpha particle as a constituent. Their high momentum, however, meant that they could enter within the atomic volume. Their deflection could reveal much about atomic structure.

The relatively large bulk of a speeding alpha particle knocks lightweight electrons out of atoms, while its own motion continues almost undisturbed. Thomson's plum pudding model of the atom supposed that positive charges were spread diffusely throughout an atom, possibly carried by light particles such as the electron. If his idea was correct then the massive alpha particles would plough straight through the atoms.

Rutherford set about investigating atomic structure with his assistant Hans Geiger. They detected the alpha particles by the faint flashes of light that they emitted upon hitting a *scintillator*: a glass plate that had been coated with zinc sulphide. The pattern of deflections confirmed that there are intense electric fields in atoms, which deflected the alphas slightly, but their results were plagued by an incessant problem of stray scattered alphas that they could not explain.

So, in 1909, Rutherford gave his young assistant, Ernest Marsden, the task of seeing if thin wafers of gold were free from this problem.

Marsden performed the test, and informed Rutherford that most of the alpha particles passed straight through but that about one in 10,000 bounced back. Can cannonballs recoil from peas? Rutherford later remarked that it was the most incredible event that had happened in his life: 'It was as if you had fired a 15-inch shell at a piece of tissue paper and it came back and hit you.' Somewhere in the gold atoms must be concentrations of material much more massive than alpha particles. What had been an annoying problem had turned into a remarkable discovery.

Rutherford puzzled about this for several months until, in 1911, he announced his solution: all of an atom's positive charge and most of its mass are contained in a compact nucleus at the centre. The nucleus occupies less than one millionth of a millionth of the atomic volume—hence the rarity of the violent collisions—and, he supposed, the electrons are spread diffusely around outside. Rutherford computed how frequently alpha particles would be scattered through various angles and how much energy they would lose if this model of an atom was correct. During the next two years, Marsden and Geiger scattered alpha particles from a variety of substances and confirmed Rutherford's theory of the nuclear atom.

Rutherford's original calculation was remarkably direct. Positively charged alpha particles are repelled by the nucleus, and even turned back in their flightpath. It was these rare cases of bouncing back through 180 degrees that first enabled Rutherford to deduce the size of the nucleus. This is how he did it.

When the alpha particle is far away from the atom, the total energy of the alpha particle is in its motion: kinetic energy. As the alpha approaches the atomic nucleus of the target, it is slowed: 'like charges repel'. Thus its kinetic energy falls, but energy overall remains constant because the potential energy due to electrostatic repulsion grows as the alpha and nucleus get near to one another. Then the alpha comes to rest momentarily before the force of electrical repulsion ejects it back from whence it came. At this point, when it is at rest, its energy is totally electrostatic potential energy, which is proportional to the product of the electrical charges and inversely proportional to the distance of approach to the nucleus.

By energy conservation, Rutherford equated this quantity, which depended on distance, to the kinetic energy of the alpha particle early in its flight, and deduced a measure of the size of the nucleus. (He had measured how far alphas travel through various materials before losing all their momentum. From this range he was able to estimate the amount of kinetic energy they had set out with).

The result astonished him. In his notes his excitement was clear, for his handwriting became almost illegible as he wrote 'it is seen that the charged centre is very small compared with the radius of the atom'.

In this we have essentially the picture of the nuclear atom that has survived for the subsequent century: negatively charged electrons and a compact positive nucleus in perfect electrical balance make a neutral atom. By contrast, in mass it is no contest because the positive nuclei outweigh the negative electrons by several thousands. Our matter is composed of 'Brobdingnagian' positives and 'Lilliputian' negatives; negative and positive electrical charges balance neatly but in a very lopsided asymmetrical fashion.

While this left no doubt that the positive charge is situated at the centre, there was still a puzzle about what the flighty electrons do. In 1913, this led to Niels Bohr's 'planetary' model of the atom, where electrical forces held the electrons remote from the nucleus. An atom is thus very empty in terms of particles but full of intense electric fields. This has led to an oft-quoted analogy that an atom's structure is similar to the solar system, the essential differences being the overall scale and that there is electromagnetic instead of gravitational attraction. However, this is a poor analogy for several reasons, one being that in reality the atom is far emptier than the solar system.

In the solar system, our distance from the Sun is 100 times larger than the diameter of the Sun itself; the atom is far emptier, with a factor of 10,000 as the corresponding ratio between the radius of an atom of hydrogen and the extent of its central nucleus—the proton. And this emptiness continues. Individual protons are in turn made of yet smaller particles—the quarks—whose intrinsic size is smaller than we can yet measure. All that we can say for sure is that is that a single quark is no bigger than 1/10,000 the diameter of a proton. The same is true for the 'planetary' electron relative to the proton 'sun': 1/10,000 rather than the 'mere' 1/100 of the real solar system. So the world within the atom is incredibly empty.

So how does matter appear to be so solid? The atom may be empty as far as its particle content is concerned, but it is filled with powerful forces. The electric and magnetic fields within an atom are far stronger than any we can make with even the most powerful magnets. It is these electromagnetic fields of force that make the atom so impenetrable and which prevent you sinking to the centre of the earth as you read this.

This picture of the nuclear atom still survives a century after it was first envisaged. If an atom were enlarged to the size of a cathedral, its nucleus would be no bigger than a fly. Nonetheless, while electrons to this day appear to be basic fundamental particles, atomic nuclei are not. An atomic nucleus has a rich structure of its own. The first steps in unravelling its nature came with Rutherford and his colleagues in the early decades of the 20th century.

Chapter 2
Nuclear alchemy

Proton—the carrier of nuclear charge

By 1913, Rutherford had established that the positively charged atomic nucleus is the source of intense electric forces, which are felt throughout the atom and which hold the negatively charged electrons in place. He realized that not only do the nuclei of all atomic elements contain electric charge but also heavy elements such as iron or gold have more than their lighter counterparts such as hydrogen, carbon, or nitrogen. An atom of gold, for example, contains 79 electrons, whose negative charge is balanced by an equally large positive charge on the compact nucleus. This far exceeds the charge on an alpha particle, which is why an encroaching alpha particle is repelled before it reaches the nucleus. Thus Rutherford's experiments had penetrated the atom, and exposed the existence of a massive compact heart, but revealed nothing about the structure of the nucleus itself.

He realized that for light atoms, with only a small electric charge on the nucleus, the alphas could make a closer approach. Hydrogen is the lightest element of all, with only one electron, thus its nucleus should be the smallest and least repulsive to alpha particles. With Marsden, Rutherford now set about firing alpha particles at hydrogen atoms.

An alpha particle is about four times heavier than the nucleus of a hydrogen atom. So when alpha particles are fired at hydrogen, one would have a situation akin to a football—the alpha particle—hitting a lightweight tennis ball—hydrogen. In such a case, the football tends to continue onwards, knocking the tennis ball forwards in the same general direction. A similar thing happens when a relatively massive alpha particle hits the nucleus of a hydrogen atom: the hydrogen nucleus is ejected forwards.

The unravelling of nuclear structure was greatly aided by the invention of the cloud chamber, which reveals the tracks of electrically charged particles as they pass through supersaturated vapour. Rutherford famously remarked that this gave science a 'telescope to look inside the atom'. It was by means of a cloud chamber that he and Marsden detected the trails of positively charged particles, with masses about one-fourth that of an alpha particle. Rutherford argued that these must be the nuclei of hydrogen atoms. Today we call these particles, each with a single unit of positive charge, protons.

His insight that protons are common to the nuclei of all atomic elements, and as such carry the positive charges of atomic nuclei, came between 1914 and 1917. Initially, Marsden had noticed that alpha particles knocked protons from atoms in the air. Rutherford then managed to eject protons from the atoms of six light elements: boron, fluorine, sodium, aluminium, phosphorous, and nitrogen. As a result, in 1919, he named these particles protons, from the Greek for 'first', as they were the first identified constituent of atomic nuclei of all elements.

Between 1921 and 1924, Patrick Blackett took 23,000 photographs of alpha particles bombarding nitrogen in a cloud chamber. In most of these images, the alphas passed clean through, or were occasionally deflected slightly like billiard balls in flight. In eight precious examples, however, something remarkable was seen.

In these, an alpha particle would hit an atom of nitrogen, and eject a proton together with something that left a short stubby track in the cloud chamber. This trail was identified as being due to the production of a moderately heavy nucleus, similar to nitrogen. Most significant, however, was that there was no sign of a recoiling alpha particle. The explanation is that the incident alpha had chipped a proton from the nitrogen, and then itself become bound to the target nucleus to form a nucleus of a form of oxygen. The alpha particle had modified the nitrogen nucleus; nuclear transmutation had been captured on film.

We can summarize this process as follows:

$$\,^4_2\,\alpha + \,^{14}_7\,N \rightarrow \,^1_1\,p + \,^{17}_8\,O$$

where the subscript is the number of protons, which defines the element (also shown by the symbols N for nitrogen, O for oxygen, with the alpha particle and proton explicit). The superscript is effectively the mass relative to the proton. (As we shall see in the next section, this is the total number of protons and neutrons.) The simple rearrangement of these constituents in the reaction is seen by the conservation of mass (superscript) and positive charge (subscript). These quantities are conserved at 18 and 9 respectively on each side of the reaction—the 'before' and 'after' of the collision that reconfigured them.

That the nuclei of all atoms are made from common components was by now clear. Transmutation occurs when those components are rearranged. However, the idea that these constituents were solely protons did not work.

The nucleus contains most of the atom's mass and all of its positive charge. Thus a nucleus with twice the charge of another should have twice as many protons and hence, supposedly, double the mass. However, this is not what happens; most atomic nuclei tend to be significantly more massive than this. For example, while eight protons in an oxygen nucleus explained its positive charge, measurements of the relative atomic masses of the chemical elements imply that oxygen atoms are about 16 or 17 times more massive than hydrogen. Oxygen's eight protons provide only half of the mass: what contributes the remainder?

The neutron

In 1920, to explain this mismatch between charge and mass number, Rutherford speculated that there was an electrically neutral analogue of the proton: the neutron. The simplest guess was that a proton and an electron somehow grip one another tightly inside the nucleus, playing the role of an effectively neutral particle.

However, this doesn't explain all the facts. The neutron is a single particle, similar to a proton. This is how its existence was first predicted, and then confirmed.

Atomic nuclei can spin, but only at certain specific rates, according to quantum mechanics. The rate of spin can be deduced from the spectrum of light emitted by a nucleus in a magnetic field, but this involves details of quantum mechanics beyond the scope of this book. For our purposes, it is the result that matters: in the case of nitrogen, for example, the results could only be explained if its nucleus contained an even number of constituents in all. The proton–electron model would have required fourteen protons to explain the mass of nitrogen, and seven electrons to give the net charge, which totals twenty-one particles, an odd number, inconsistent with the even number required by the spin measurement. However, if instead of a proton and electron joining together, there is a single genuine particle, as heavy as a proton but with no electrical charge—a neutron—everything fits. Replacing the seven proton and electron pairs by seven neutrons gives the same charge and mass as before but now involves a total of fourteen particles, an even number as required by the spin.

The spin of nitrogen is satisfactorily described if the neutron spins at a rate identical to the proton. The spins of other elements also fit with the hypothesis. So the idea is economical and precise: all atomic nuclei are built from protons and neutrons.

The challenge in the 1920s was to prove the idea, and the first step would be to demonstrate the neutron's existence. Frédéric Joliot, and his wife Irene Curie, daughter of Marie Curie, had evidence for the neutron, but misinterpreted it. The neutron was discovered in 1932 by Rutherford's colleague, James Chadwick. This is how it came about.

The Joliot-Curies had fired alpha particles at beryllium, the fourth lightest element, and discovered that an electrically neutral radiation came out. Today, we know that these were neutrons; however, they mistakenly thought them to be X-rays. When Rutherford heard of their results he realized that they had probably inadvertently produced neutrons, and Chadwick confirmed this by performing a nuclear analogue of a canon shot at snooker.

The alphas hit the beryllium and ejected the mystery radiation; this radiation then hit the atoms in a variety of other gases—hydrogen, helium, and nitrogen. The lightweight hydrogen recoiled a lot, whereas the heavier elements recoiled less, a pattern consistent with the invisible agent having a mass similar to that of a proton. Chadwick then placed a range of elements in the path of the neutral particles. They ejected protons from each and every element that he tried, even from the gold in Rutherford's Nobel Prize medal. In each case, the energy of the emerging protons was only consistent with them having been ejected by a massive neutral particle.

Rutherford compared this to H.G. Wells' invisible man: although you could not see him directly, his presence could be detected when he collided with people in the crowd. Here was evidence for would come from experiments by Enrico Fermi's group in Rome in 1934, which imitated the Joliot-Curies' experiments but with neutrons in place of alpha particles, and the discovery of nuclear fission by Hahn and Strassmann in Germany in 1938.

Fermi realized that because neutrons have no electrical charge, they are not repelled as they approach a nucleus, and so they can more easily gain access than do alpha particles or protons.

To prevent the neutrons hitting the nucleus hard and shattering it, Fermi first slowed the neutrons by passing them through paraffin or water. With this technique he successfully modified the nuclei of various atoms. He attached neutrons to fluorine, producing a new artificial isotope of that element and did likewise with a total of 42 different nuclear targets until he came to the heaviest known element, uranium.

Irradiating uranium gave him some puzzling results, which suggested that the neutron had not simply become attached to the uranium nucleus. Fermi thought that he had produced the first *transuranic* element, one place above uranium in Mendeleev's table, unknown at that time on earth but capable of existence in principle. In reality he had split uranium in half, but he did not realize it.

In Germany, Otto Hahn and Fritz Strassmann made a similar experiment, and identified barium among the products. Their former colleague, Lisa Meitner, and her nephew, physicist Otto Frisch, came up with the explanation. We will meet nuclear forces in chapter 3, but in summary, a uranium nucleus is like a liquid drop. Liquid drops are held together by surface tension, while the nucleus is held by the strong force. The electrical repulsion among the protons in a nucleus works against the strong force; the heavier the element, and the more numerous the protons, the bigger is the repulsion. Beyond uranium the two forces work against each other and cancel, such that no stable elements occur.

Uranium itself is so delicately balanced that slow neutron bombardment makes a uranium nucleus wobble like a liquid drop and break up. There are several possible end products, one example of which is

$$n + {}^{235}_{92}U \rightarrow {}^{144}_{56}Ba + {}^{89}_{36}Kr + 3n$$

The splitting of nuclei in this way is called nuclear fission.

The amount of energy released in fission is much greater than in radioactive decays, whether natural or induced. Furthermore, the production of three 'new' neutrons among the fragments can trigger the breakup of other uranium nuclei, and enormous energies can be released in the ensuing *chain reaction*. This is the basis of practical nuclear power and of the so-called atomic bomb.

Energy, waves, and resolution

As we said earlier, Rutherford discovered the existence of the atomic nucleus by using beams of alpha particles, but he was unable to resolve the internal structure of the nucleus by that means. To understand the reason why, and what is required to make experimental investigations of the nucleus, we need a small diversion on the link between distance, energy, and resolution.

To find out what something is made of you might use light to look at it, heat it and see what happens, or smash it by brute force. The common feature is energy. In the latter pair of cases energy's role is immediately apparent; in the case of light the link is less obvious but the link comes courtesy of quantum theory.

Light is a form of electromagnetic radiation. As you go from red light to blue, the wavelength halves, the wavelength of blue light being half that of red (or equivalently, the frequency with which the electric and magnetic fields oscillate back and forth is twice as a new subatomic particle, similar in mass to the proton but with no electrical charge: the neutron that Rutherford had predicted. Apart from a one part in a thousand difference in mass and the presence of electrical charge, the proton and neutron are identical. As they are constituents of the nucleus, they are often referred to collectively as *nucleons*.

In 1932, the same year that Chadwick discovered the neutron, nuclei were split for the first time by *artificial* means. Whereas previously alpha particles produced by the *natural* radioactive decays of radium had been used as probes, now John Cockcroft and Ernest Walton used electric fields to accelerate protons to high speed, and then fired these beams of high energy particles at lithium nuclei.

This had two advantages over what had been done before. First, protons with their single positive charge feel less electrical resistance than do the doubly charged alpha particles when approaching a nucleus. Second, and more important, the high-speed particles penetrate deeper before being slowed. Cockcroft and Walton had thus made the first nuclear particle accelerator, and created a practical tool for investigating within the atomic nucleus. This device, colloquially referred to as an *atom smasher*, was the prototype of the modern particle accelerators that have been used for probing the internal structure of the neutrons and protons themselves. It was from 1932 onwards that the structure of atomic nuclei began to be decoded.

Isotopes

Like protons but with no electric charge, neutrons are present in nuclei, and add to the nuclear mass without changing the total charge. The neutron is an essential component of all atomic nuclei, save that of hydrogen, which normally consists of just a single proton.

Every nucleus of a given element contains the same number of protons but may have different numbers of neutrons. Hydrogen usually has one proton and no neutrons but about 0.015% of all hydrogen atoms consist of a proton and a neutron. This is known as the *deuteron*, the nuclear seed of deuterium, sometimes called *heavy hydrogen*. A proton accompanied by two neutrons is known as the *triton*, which is the nuclear seed of tritium.

A rare form of uranium, known as U-235, and the common form, known as U-238, have 143 and 146 neutrons respectively which, added to the 92 protons in each case, gives a total number of 235 or 238 constituents. It is traditional to denote atomic nuclei by a symbol corresponding to the element; the number of protons is shown as a subscript and the number of nucleons (neutrons plus protons) as a superscript. Thus for example the alpha particle may be denoted by $^{4}_{2}He$ while $^{235}_{92}U$ and $^{238}_{92}U$ are the two particular isotopes of uranium just mentioned. This code is useful for keeping account of how the individual neutrons and protons are exchanged during nuclear reactions—you just count the numbers in the superscripts (or numerator) and subscripts (denominator). The total number of nucleons (numerator) is conserved but the denominator may change, so long as electric charge is conserved (for example, by emission of a beta particle, that is an electron, as we shall see later). The various nucleons are shown in Figure 2.

Although such isotopes have the same chemical identity, the behaviour of the atomic nucleus itself can change dramatically. Indeed, the neutron number is key to extracting energy from the nucleus, either gradually in nuclear reactors or explosively in weapons. For example, U-235, forms the raw material for both nuclear power and atomic bombs.

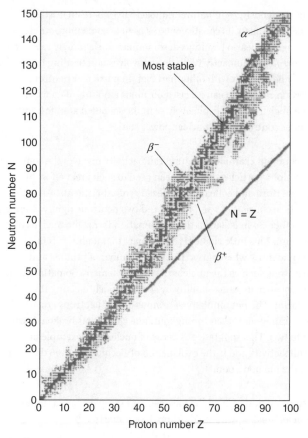

2. A chart of the nuclei. The solid black entries are the most stable isotopes. Grey entries, which are unstable, may decay either by β^- or β^+ emission, or by α emission as shown.

The sources of radioactivity

Protons and neutrons are the common ingredients of *all* nuclei and so one variety of nucleus can transmute into another by absorbing or emitting these particles. This may happen spontaneously, as in natural radioactivity, where an unstable nucleus adjusts its constituents, so as to become more stable. Such changes may also be induced, sometimes misleadingly referred to as *artificial* radioactivity. To do so involves bombarding the nuclei of some element with other particles, in particular neutrons. In certain cases, this can release enormous amounts of energy from the nucleus, either explosively as in the so-called atomic bomb, or under control as in a nuclear power station.

A common example of natural radioactivity involves the source of the alpha particles that had been used to such great effect by Rutherford and colleagues in decades of experiments. An alpha particle consists of two protons and two neutrons tightly clumped together. So in isolation this combination forms the nucleus of helium. This little cluster is so compact that it almost retains its own identity when buried in a large number of protons and neutrons such as the nucleus of a heavy element. Sometimes, the heavy nucleus gains stability by spontaneously ejecting the quartet. The net numbers of protons and of neutrons are separately conserved throughout: one cluster has broken down into two. This spontaneous decay of nuclei is an example of radioactivity and is the explanation of Becquerel's 1896 discovery of the phenomenon.

The emission of an alpha particle, for example when uranium nuclei break down into thorium, is summarized by:

$$^{238}_{92}U \rightarrow \,^{234}_{90}Th + \,^{4}_{2}He$$

Isotopes that contain a high percentage of neutrons tend to be unstable (see chapter 3). Not only can they emit alpha particles, but a neutron can spontaneously convert to a proton, and emit an electron in the process. Electrical charge is conserved:

$$n^0 \rightarrow p^+ + e^-$$

The superscripts here denote the electric charges of the n (neutron), p (proton), and e (electron).

This process is known as β (beta) radioactivity, and is the source of many nuclear transmutations. One puzzling feature was that in beta decay the electron's energy was seen to vary from one occasion to another, whereas it should have been the same every time, if the process was as described previously. Energy is conserved over long timescales but can be converted from one form into another. Einstein showed that an amount of energy $E = mc^2$ is produced when a mass m is destroyed, where c is the velocity of light. An isolated neutron has slightly more mass than a proton; in energy this corresponds to some 1.3 MeV in 940 MeV. (The explicit definition of these energy units will be given later.) Thus the energy released in the decay should be

$$E = m(neutron)c^2 - [m(proton)c^2 + m(electron)c^2] = 0.8\,\mathrm{MeV}$$

which should be manifested as kinetic energy of the proton and electron. However, far from being fixed at this value, in practice the electron energy varies all the way down to zero.

In 1931, Austrian Wolfgang Pauli proposed that some unseen third particle was also being produced in the decay. This invisible particle takes away some of the energy itself, which explains the variability of the electron's energy. This particle is electrically neutral, and was named the *neutrino* (to distinguish it from the neutron). It is conventionally denoted by the symbol ν. Thus beta decay in full, as shown in Figure 3, is as follows:

$$n^0 \rightarrow p^+ + e^- + \nu$$

The existence of the neutrino was not confirmed until 1956. (Technically this process produces an antineutrino, $\bar{\nu}$, but this detail goes beyond the range of the present book.) Neutrinos are ejected in nuclear transmutation but have no existence

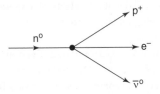

3. Neutron beta decay. A neutron, denoted n^0, turns into a proton p^+, an electron e^-, and antineutrino $\bar{\nu}$; the superscripts denote the amount of electric charge that each particle has relative to that of a proton, and the sign denotes whether it is positive or negative.

within a nucleus. As such they are like the electron produced by β decay.

Finally we come to gamma radioactivity.

Quantum mechanics explains the behaviour of electrons in atoms, and of nucleons in nuclei. In an atom, electrons cannot go just where they please, but are restricted like someone on a ladder who can only step on individual rungs. When an electron drops from a rung with high energy to one that is lower down, the excess energy is carried away by a photon of light. The spectrum of these photons reveals the pattern of energy levels within the atom. Similar constraints apply to nucleons in nuclei. Nuclei in *excited* states, with one or more protons or neutrons on a high rung, also give up energy by emitting photons. The main difference between what happens to atomic electrons relative to atomic nuclei is the nature of the radiated light. In the former the light may be in the visible spectrum, whose photons have relatively low energy, whereas in the case of nuclei the light consists of X-rays and gamma rays, whose photons have energies that are millions of times greater. This is the origin of gamma radioactivity.

Nuclear energy scales and units

At this point it is worth summarizing how energies are accounted for in nuclear physics.

fast for blue light as red). The electromagnetic spectrum extends further in both directions. Beyond the blue horizon—where we find ultraviolet, X-rays, and gamma rays—the wavelengths are smaller than in the visible rainbow; by contrast, at longer wavelengths and in the opposite direction, beyond the red, we have infrared, microwaves, and radio waves. Our inability to see atoms has to do with the fact that light acts like a wave, and waves do not scatter easily from small objects. To see a thing, the wavelength of the beam must be smaller than it is.

X-rays are light with such short wavelengths that they can be scattered by regular structures on the molecular scale, such as are found in crystals. The wavelength of X-rays is larger than the size of individual atoms, so the atoms are still invisible. However, the distance between adjacent planes in the regular matrix within crystals is similar to the X-ray wavelength and so X-rays begin to discern the relative position of things within crystals. This technique is known as X-ray crystallography.

One of the great discoveries in the quantum theory was that particles can have a wavelike character, and conversely that waves can act like staccato bundles of particles, known as quanta. Thus an electromagnetic wave acts like a burst of quanta, or photons. The energy of any individual photon is proportional to the frequency (v) of the oscillating electric and magnetic fields of the wave. This is expressed in the form

$$E = h v$$

where the constant of proportion, h, is Planck's constant. The length of a wave (λ) and the frequency with which peaks pass a given point, are related to its speed, c, by $v = c / \lambda$. So we can relate energy and wavelength

$$E = \frac{hc}{\lambda}$$

and the proportionality constant $hc \sim 10^{-6} \, \text{eVm}$. This enables us to relate energy and wavelength by the approximate rule of thumb: '1 eV corresponds to $10^{-6} \, \text{m}$', and so on.

To probe deep within atoms and reveal nuclear structure we need a source of very short wavelength, or equivalently, of high momentum and energy. We can look at as small a distance as we like; all we have to do is to speed the particles up, give them more and more energy to get to ever smaller wavelengths. To resolve distances on the scale of the atomic nucleus, $10^{-15} \, \text{m}$, requires energies of the order of GeV.

Naturally occurring alpha particles don't pack much punch. They are ejected from heavy nuclei with only a few MeV of kinetic energy, or equivalently a few MeV/c momentum, and as such are able to resolve structures on distance scales larger than about $10^{-12} \, \text{m}$. Now, such sizes are smaller than those of atoms, which makes such alphas so useful, but are still much larger than the $10^{-14} \, \text{m}$ extent of even a large nucleus, such as that of a gold atom, let alone the $10^{-15} \, \text{m}$ size of the individual protons and neutrons that combine to make that nucleus. So although such alphas were fine for discovering the existence of the atomic nucleus, to see inside such nuclei would require beams with more energy.

It was in 1932 that the first accelerator of electrically charged particles was built in Cambridge by John Cockcroft and Ernest Walton, and a detailed picture of nuclear structure, and of the particles that build it, began to emerge. One can use beams of atomic nuclei, but while these were truly *atom* (or rather *nucleus*) *smashers*, and helped to determine the pattern of nuclear isotopes and their details, the clearest information on their basic constituents came with the simplest beams. A nucleus of carbon contains typically six protons and a similar number of neutrons. As such there is a lot of debris when it hits another nucleus, some coming from the carbon beam itself as well as that from the target. This makes a clear interpretation difficult. The signals are easier to interpret when one uses a beam of just protons or electrons; these are the main ways of probing the nucleus, and distances down to 10^{-19} m today.

Chapter 3
Powerful forces

A strong attraction

Why do atomic nuclei exist at all? A large nucleus contains many protons in close proximity to one another. Why do these protons, all with the same electrical charge, not mutually repel?

The answer is that there is a strong attractive force that acts between neutrons and protons when they are in contact with one another. This force does not distinguish between them: neutrons and protons attract one another with the same strength as either attracts its own kind. Within the nucleus, where protons and neutrons are in close proximity, this strong attraction is over a hundred times more powerful than the electrical repulsion.

There is a limit, however, to the number of protons that can co-exist like this. For any individual proton, the attractive glue acts only between it and its immediate neighbours, but electrical disruption acts across the entire volume of the group. In a large nucleus, the total amount of electric repulsion can exceed the localized attraction, and the nucleus cannot survive. The neutron, being electrically neutral, feels no such disruption. Thus the presence of neutrons can stabilize the nucleus. Even so, it takes many neutrons to do this, especially in larger nuclei.

This is why heavy nuclei tend to have more neutrons than protons. The further we go up the periodic table of the atomic elements, the larger are the nuclei, and the greater is the excess of neutrons required to maintain stability. Too great an excess of neutrons, however, will destabilize a nucleus. The reason is that a neutron is slightly heavier than a proton, and so by Einstein's equivalence between mass and energy, a neutron has slightly more energy locked within it than there is in a proton. This extra energy leads to instability, so much so that an isolated neutron has a half-life of only about ten minutes, whereas a single proton can exist for aeons, and possibly even forever.

An isolated neutron at rest contains an amount of energy $E = m_n c^2$, where m_n is the neutron mass. An isolated proton likewise would possess an amount $E = m_p c^2$, which empirically is slightly less than that of the neutron. So when an isolated neutron undergoes beta decay $n \to pe^-\bar{\nu}$ it loses energy—which is carried away by the electron and neutrino—and ends up in a lower energy state: a proton. The proton is the lightest nucleon and cannot lower its energy by turning into anything else. As a result an isolated proton is stable.

Now imagine that neutron in a nucleus, which contains also one or more protons. The simplest example is the deuteron, a form of the hydrogen nucleus consisting of a single neutron and a single proton. This is stable. Why is the neutron stabilized here?

To answer this, think what would be the end result were the neutron to decay. At the start, we had a neutron and proton touching; at the end there would be two protons. So although the neutron would have lost energy in its conversion to a proton, the two protons that ensue would have a mutual electrical repulsion, which adds to the energy accounts. The amount of this electrostatic increase exceeds that lost in the $n \to p$ conversion. Overall, the net energy would increase in going from the $(np) \to (pp)$ pair. Thus in this case the beta decay of the neutron would increase the total energy of the nuclear particles, and so the initial (np) is stable and survives.

This generalizes to nuclei where there are similar numbers of neutrons and protons: beta decay is prevented due to the extra electrical effects that would occur in the final 'proton rich' environment. However, try putting too many neutrons in one ball and eventually the downhill energy advantage of them shedding a little bit of mass on turning into protons is more than the uphill electrical disadvantage of the protons' electrical repulsion. So there is a limit to the neutron excess in nature.

The energy accounts that are at work in the neutron stability example also explain the phenomenon of 'inverse beta decay' or 'positron emission'. There are some nuclei where a proton can convert into a neutron, emitting a positive version of the electron (known as a positron, the simplest example of an antiparticle). This basic process is represented as $p \rightarrow ne^+\nu$.

This does not happen for a free proton as it would have to 'go uphill', in that the energy $m_n c^2$ of the final neutron would exceed that of the initial proton due to its extra mass. However, in a nucleus where there are several protons, this inverse beta decay can *reduce* the electrostatic energy as there will be one less proton at the end than at the start. If this reduction in energy exceeds the price of replacing a $m_p c^2$ by the larger $m_n c^2$, then inverse beta decay, $A(P,N) \rightarrow A(P-1, N+1) e^+\nu$ will occur, where A denotes the total number of neutrons and protons. There are several examples of natural 'positron emitters' in nature. They have great use in medicine. The PET scanner, positron emission tomography, exploits this phenomenon.

The result of all this is that if there are too many neutrons in a nucleus, the assembly becomes unstable. Only a limited number of stable isotopes exist, namely those where the number of neutrons is similar to, or somewhat larger than, the number of protons. If a nucleus is created with too many neutrons—in high-energy collisions for example, or within a supernova—it will become more stable by the process of beta decay, in which a neutron converts to a proton, at the same time emitting an electron (the beta particle) and a neutrino. A collection of more than about 10^{57} neutrons can hold together under gravity and form a neutron star—this will be described in chapter 6.

It is possible that you are wondering: if the attractive force between protons and neutrons is the same, why don't we have a bound state of two neutrons, which can beta decay into a deuteron (proton and neutron)? The answer depends on effects of quantum mechanics that, in summary, are as follows. The strengths of the attractions are almost the same, but depend also on the relative orientation of the spins of the constituents. The constraints of quantum mechanics require that two such neutrons cannot spin simultaneously in the same orientation—in the jargon, they must combine to a state with no total spin. The result is that the pair is almost bound, but not quite. We shall discuss this further when we meet exotic halo nuclei in chapter 7.

In atoms, electrons arrange themselves in quantum levels called shells. This also happens with protons and neutrons in nuclei, where these nucleons arrange themselves following similar rules. In atoms, the shell with the lowest energy is full when it contains two electrons. In nuclei it is full when it contains two neutrons and two protons. This makes an exceptionally stable configuration—the nucleus of helium, the alpha particle. As we noted earlier, it is this stability of alpha particles that accounts for their appearance in radioactive decays. This is especially true for heavy elements such as thorium or uranium, which become lighter, and more stable, by shedding protons and neutrons in alpha particle clusters. This is described in chapter 5.

From pions to quarks

Shake electrons and they emit electromagnetic radiation as the electric field surrounding the charged particle is disrupted. Pummel protons or neutrons and a burst of radiation is released, which consists of particles called pions. Pions are emitted when the nuclear force field is disturbed. The more violent the disturbance, the greater is the number of pions that emerge.

Pions occur with either positive or negative electric charge, or with none at all. They are denoted respectively: π^+, π^-, π^0. The mass of a pion is about 15% that of a proton or neutron, or about 140 MeV in the units that are traditionally used in nuclear physics.

In quantum theory, forces are transmitted by the exchange of particles. A model of the strong nuclear force, when it acts between neutrons and protons in an atomic nucleus, is that it is transmitted by pions. Quantum uncertainty allows such a particle to exist fleetingly, at no energy cost, so long as the total time is restricted to less than an amount governed by Planck's constant, denoted h:

$$\Delta T \leq \frac{h}{4\pi \Delta E}$$

where ΔT is the time allowed when an amount of energy ΔE is 'overdrawn' from the accounts.

Thus as a photon has no mass, and can, in extremis, carry no energy, there is no need to borrow any energy: $\Delta E = 0$ and so $\Delta T = \infty$, whereby the electromagnetic force has potentially infinite range. However, a free pion has mass, and so an energy of at least 140 MeV. For two nucleons to exchange a pion within a nucleus, the pion has to pop into existence at no cost of energy. Thus it has to 'borrow' 140 MeV, which quantum uncertainty limits to about 10^{-23} seconds. In this time it travel no more than about 10^{-15} metres, a femtometre, which thus limits its range of action. This is the reason why the resulting strong force acts over such a short distance.

The existence of the pion was predicted by Japanese theorist, Hideki Yukawa, in 1935, with this very role in mind. It was eventually discovered, in 1947, in cosmic rays. The primary rays consist of protons and atomic nuclei, and collisions between these and atoms in the upper atmosphere spawned pions, whose trails were detected in photographic emulsions.

Although pions were proposed as transmitters of the strong nuclear force, and theoretical models of this force use this hypothesis even today, pions are not classified with other force-transmitting particles such as electromagnetism's photon, or the W and Z bosons, which transmit the weak force, such as manifested in beta decay for example. This is because pions are now known not to be elementary particles but composites made of *quarks*. This is true of protons and neutrons also. As Rutherford discerned the nucleus at the heart of the atom by the scattering of low energy alpha particles, so the scattering of very high energy beams of electrons, in the 1960s and 1970s, revealed that there is a deeper level of structure with protons and neutrons: all particles that feel the strong interaction are made of quarks.

The colour force and QCD

Quarks that form nuclear particles come in two *flavours*, known as up (u) or down (d), with electrical charges that are fractions, +2/3 or −1/3 respectively, of a proton's charge. Thus uud forms a proton and ddu a neutron. In addition to electrical charge, quarks possess another form of charge, known as *colour*. This is the fundamental source of the strong nuclear force. Whereas electric charge occurs in some positive or negative numerical amount, for colour charge there are three distinct varieties of each. These are referred to as red, green, or blue, by analogy with colours, but are just names and have no deeper significance.

The triplication apart, colour charge and electric charge obey very similar rules. For example, analogous to the behaviour of electric charge, colour charges of the same colour repel, whereas different colours can attract (technically, when in an antisymmetric quantum state). A proton or neutron is thus formed when three quarks, each with a different colour, mutually attract one another. In this configuration the colour forces have neutralized, analogous to the way that positive and negative charges neutralize within an atom. As atoms form molecules, due to the electrically charged constituents within them, so neutrons and protons form atomic nuclei due to the coloured constituents within nucleons. Thus atomic nuclei can be thought of as like 'molecules of colour charge'.

The relativistic quantum theory of colour is known as quantum chromodynamics (QCD). It is similar in spirit to quantum electrodynamics (QED). QED implies that the electromagnetic force is transmitted by the exchange of massless photons; by analogy, in QCD the force between quarks, within nucleons, is due to the exchange of massless *gluons*. Not only are these bizarre particles required by QCD theory, but there is experimental evidence for their reality.

One of the remarkable implications of QCD theory, which is confirmed experimentally, is the manner in which the forces between quarks and between nucleons behaves at different distances. Whereas the force between two quarks is relatively feeble when they are within less than 10^{-16} m of one another, when separated by 10^{-15} m or more, its effects become very strong. In such circumstances the potential energy exceeds hundreds of MeV. In such a case, new quarks and antiquarks materialize in the force field. As a result a quark and an antiquark, each with the same colour charge but of opposite sign—particle and antiparticle carrying positive and negative charges, say—mutually attract and can form a pion. Thus modern theory implies that the exchange of pions between nucleons is the manifestation at nuclear scales of a more fundamental property—the existence of quarks and the forces between coloured charges as described by QCD.

While this explains the fundamental source of the strong nuclear force as being due to quarks, it is nonetheless for practical purposes useful to describe the strong force, when acting over nuclear distances, as due to the exchange of pions.

Quarks in nuclei

The scattering of beams of high energy electrons from nucleons revealed the presence of quarks within the target particles. The data reveal directly the spread of quarks in momentum, and by quantum theory—where the uncertainty or spread in momentum and position are complementary—this may be linked to their spatial confinement. All the data are consistent with the quarks being effectively free yet confined within a nucleon of about 1 femtometre radius. This is true for free protons, or for protons and neutrons within a loosely bound nucleus such as the deuteron of heavy hydrogen. However, when similar experiments were made with the targets being heavy nuclei, such as iron, the quarks showed a subtle, yet significant, difference in behaviour. This became known as the *EMC effect*, after the European Muon Collaboration at CERN, who first discovered the phenomenon around 1980.

In a nutshell, the quarks in heavy nuclei are found to have, on average, slightly lower momenta than in isolated protons or neutrons. In spatial terms, this equates with the interpretation that individual quarks are, on average, less confined than in free nucleons. This phenomenon has been studied over three decades, using beams of muons (in effect heavy analogues of electrons), electrons over a wide range of energies, and neutrinos. The overall conclusion is that the quarks are more liberated in nuclei when in a region of relatively high density. Thus the effect is more marked for heavy nuclei than for lighter ones, with an increased prominence for dense nuclei.

Perhaps this is not totally surprising. As nucleons are bound in nuclei by the exchange of pions, and as these pions contain quarks, and ferry them from one nucleon to another in the tightly packed nucleus, there is a chance for a quark to experience greater freedom than when locked within a 1 femtometre prison, such as when in the proton of hydrogen say. While this is doubtless a substantial part of the phenomenon, the data also suggest that there is a small probability for quarks to find themselves in localized clusters of six, rather than in the separate groups of three that we recognize as nucleons.

This interpretation of the microstructure of atomic nuclei suggests that nuclei are more than simply individual nucleons bound by the strong force. There is a tendency, under extreme pressure or density, for them to merge, their constituent quarks freed to flow more liberally.

This freeing of quarks is a liberation of colour charges, and in theory should happen for gluons also. Thus, it is a precursor of what is hypothesized to occur within atomic nuclei under conditions of extreme temperature and pressure. Here we have another potential analogy between the familiar world of electric charge and more exotic one of colour. For example, atoms are unable to survive at high temperatures and pressure, as in the sun for example, and their constituent electric charges—electrons and protons—flow independently as electrically charged gases. This is a state of matter known as *plasma*. Analogously, under even more extreme conditions, the coloured quarks are unable to configure into individual neutrons and protons. Instead, the quarks and gluons are theorized to flow freely as a *quark–gluon plasma* (QGP).

Quark–gluon plasma

Quarks and gluons first emerged from the heat energy of the big bang. Whereas in today's cold universe these fundamental particles are trapped inside protons and neutrons, in the initial heat and pressure, quarks and gluons would not have stuck to one another in these identifiable clusters. Instead, they would have existed in the dense energetic 'soup' known as quark–gluon plasma. Theoretical understanding of the formation of matter in the big bang posits that QGP was the primordial state of matter during the first few millionths of a second, just before it 'froze' into particles such as protons, neutrons, and pions.

To simulate these conditions in experiments today, heavy nuclei such as gold or lead are smashed into one another at high energy. At CERN in the 1990s, beams of heavy nuclei were fired at static targets of heavy elements. In 2000, the Relativistic Heavy Ion Collider (RHIC) at Brookhaven National Laboratory in the USA began the first dedicated collisions of counter-rotating beams of heavy nuclei. As with simpler particles, such as electrons and protons, the great advantage of a colliding beam machine is that all the energy gained in accelerating the particles goes into the collision. Since 2009, the blue riband of energy has been taken by the Large Hadron Collider (LHC) at CERN, where hundreds of protons and neutrons in each nucleus crash into each other at energies of over 1 TeV each. In such experiments, the protons and neutrons squeeze together, forming a fireball of quarks and gluons, at densities as much as 50 times greater than found in 'cold' nuclei on earth.

At these extreme energies, akin to those that would have been the norm in the universe when it was less than a trillionth of a second old, the nuclei 'melt'—in other words, the quarks and gluons flow throughout the region of impact rather than remaining frozen into individual neutrons and protons. At the LHC, in collisions between heavy nuclei, QGP should become commonplace, so that experimenters can study its properties in detail.

It is one thing to have created a QGP, another to demonstrate the fact. The QGP forms deep within the hot centre of the collision; the detectors, however, are relatively remote, out in the cool. By the time any debris escapes from the QGP, its constituent quarks and gluons have combined to make conventional particles such as protons, neutrons, and pions. So one has to seek some anomalies in the way these particles emerge, which are like fossil relics of the QGP and thereby show that a QGP has been formed.

Various signatures have been suggested. While conclusive proof remains controversial, the indications that QGP forms are consistent with the theory. For example, strange particles arise when an up or down variety of quark is replaced by a strange quark. The strange quark is more massive than its up or down counterparts, containing at least 100 MeV more energy than its cousins, which makes a strange quark marginally less likely to appear from a conventional collision. This is in accord with decades of experimental data. In the heat of a QGP, however, this penalty should no longer apply, in which case there should be relatively more strange particles produced from a QGP than from a conventional source. There are some hints that this is the case empirically, but the phenomenon may have other explanations and the interpretation is complex.

Another potential signal for QGP comes from *jets*. When individual protons collide head on, quarks or gluons in the protons of the two beams may smash directly into one another and bounce off back-to-back. Their energy materializes in jets of particles, such as pions, protons, and strange particles. In the case of proton collisions, the properties of these jets, such as their distribution in space, their average energies, and their particle content, have been studied for decades and agree with the underlying theory of quantum chromodynamics. Thus the observation of jets when heavy nuclei collide, and the discovery that there are significant differences of detail compared to the case when individual nucleons collide, give clues that QGP had been formed.

The data that emerge from the collisions between heavy nuclei show that in some cases one jet may be enfeebled or even extinguished entirely. This is as expected if a QGP has formed, because in such a case a jet loses some energy during its flight through the hot fireball. By studying millions of examples, information is accumulating on how the jets lose their energy, what directions they emerge in, and what particles they contain. Comparison of these changes relative to the properties of jets produced in the collisions of individual protons, or of light nuclei, or of heavy nuclei at lower energies, can eventually reveal what the fireball consists of. These experiments will be made at higher energies than hitherto at the Large Hadron Collider from 2015. Comparison of the results with those at RHIC should enable the properties of quark–gluon plasma to be established within a few years.

Chapter 4
Odds, evens, and shells

Abundance of the elements

Although the seeds of all atomic nuclei differ merely by the numbers of neutrons and protons, their abundances vary widely. Some, such as oxygen, calcium, and silicon are very common. However, there are many others that you may never have heard of, such as ruthenium, holmium, and rhodium.

As a rough rule, the ones that you think of first are among the most common while the ones that you have never heard of are the rarest. The names oxygen and carbon are in everyone's lexicon. Contrast this with astatine or francium, which amount to less than a gram of the earth's crust.

The relative abundance of different elements is something that most people are likely to agree on; the abundance of different isotopes, however, is further from common experience. Although all isotopes consist of just protons and neutrons, and differ merely in the numbers of these constituents, there is a noticeable preference for even numbers of each variety than of odd. This shows up in three distinct ways. First, there are more nuclei where the total number of nucleons A is an even number than when A is odd: 153 relative to 101. More dramatic is how the 153 members of the A even set are apportioned. As $A = Z + N$, where Z, N are the number of protons and neutrons respectively, if A is even one may have both Z and N even, or both odd. These are referred to as *ee* and *oo* respectively for 'even-even' and 'odd-odd'. Here are the numbers: the 153 isotopes with A even consist of 148 even-even and a mere five that are odd-odd. This is too big a mismatch to be mere coincidence. It is a clue to some fundamental property of the laws governing nuclear structure.

Magic numbers and shells

In chapter 2 we saw how quantum mechanics constrains the energy states of nucleons in nuclei, analogous to that of electrons in atoms. The quantum rules also limit the number of constituents that can occupy any given rung on the energy ladder. The lowest level in atoms, for example, can accept at most two electrons; an atom with three or more electrons is thus forced to have the excess on higher energy rungs. Thus in an atom of helium, with two electrons, the lowest rung—or *shell*—is full. This is a reason for the chemical stability or inertness of helium.

For nucleons the same rule applies except that now the lowest rung can accept at most two protons and two neutrons, giving a total maximum of four. Thus hydrogen has isotopes with one neutron (deuteron) or two (tritium) but not three (quadium). The third neutron, required to form quadium, would be forced to a higher energy rung, which makes the resulting isotope highly unstable. It decays immediately (half-life of a mere 10^{-22} s) by emitting a neutron and leaving tritium. Hydrogen is the only element whose isotopes are given distinct names.

Two protons could exist alone, in principle, but their mutual electrical repulsion negates this. Add one neutron, however, and the additional strong attraction stabilizes the system to give the isotope helium-3. Add another neutron, giving a total of two protons and two neutrons, and the lowest energy shell is now saturated. The result is the highly stable alpha particle, the nucleus of helium-4. Add more neutrons to helium-4 and unstable forms result. These can have interesting properties, nonetheless, forming *haloes* of two or more neutrons remote from a central core. Such *halo nuclei* are examples of several exotic clusters of neutrons and protons, which are discussed in chapter 7.

Isotopes where the shells are filled with either neutrons or protons are especially stable. The quantum rules that determine how many are allowed at each successive level are complex. In a nutshell, without explanation, an individual rung can carry twice an odd number of identical particles. Thus, in sequence, 2, 6, 10, etc. (The odd number corresponds to the following: double the amount of rotary or angular momentum, in units of Planck's constant, that the particles carry in that shell and add one, thus a total of two when the angular momentum is zero; six when it is one, etc.) The order in which rungs of various values occur depends on how the nucleons interact with one another, specifically, on the dependence of their potential energy on their spatial location. For the electron in hydrogen this varies as the inverse of the distance—the Coulomb potential–and the order of values is predictable, and explains the pattern in the periodic table of the elements. For nucleons in a nucleus, however, it is more complicated, and has to be deduced empirically from the observed pattern of the shells (Figure 4).

Qualitatively, the shells with small angular momentum, and hence low occupancy, arise first, the higher ones coming in later. Thus the first shell involves two, as we saw, whereby the alpha particle is 'doubly magic', in that it contains two protons and also two neutrons. The next shell allows six additional members, and hence eight in total. We say that eight is a *magic number* at which unusual stability arises. This is confirmed empirically because eight protons and eight neutrons, which is doubly magic, gives oxygen-16, $^{16}_{8}O$—the commonest element and isotope on our planet.The shell that now opens up beyond oxygen contains the first collection of nucleons with two units of angular momentum,

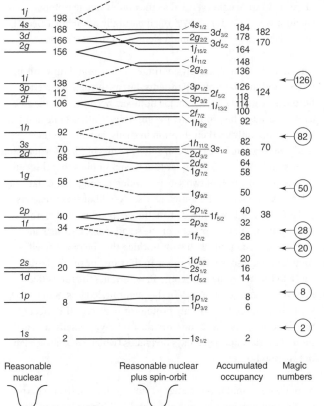

1j	198	
4s	168	
3d	166	
2g	156	

		$4s_{1/2}$ $3d_{3/2}$	184	182
		$2g_{9/2}$ $3d_{5/2}$	178	170
		$1j_{15/2}$	164	
		$1i_{11/2}$	148	
		$2g_{9/2}$	136	

1i	138	
3p	112	
2f	106	

		$3p_{1/2}$ $2f_{5/2}$	126	124	← (126)
		$3p_{3/2}$ $1i_{13/2}$	118 114		
		$2f_{7/2}$	100		
1h	92	$1h_{9/2}$	92		

3s	70	$1h_{11/2}$ $3s_{1/2}$	82	70	← (82)
2d	68	$2d_{3/2}$	68		
		$2d_{5/2}$	64		
1g	58	$1g_{7/2}$	58		
		$1g_{9/2}$	50		← (50)

2p	40	$2p_{1/2}$ $1f_{5/2}$	40	38	
1f	34	$2p_{3/2}$	32		
		$1f_{7/2}$	28		← (28)
					← (20)

2s	20	$1d_{3/2}$	20		
		$2s_{1/2}$	16		
1d		$1d_{5/2}$	14		

| 1p | 8 | $1p_{1/2}$ | 8 | | ← (8) |
| | | $1p_{3/2}$ | 6 | | |

| 1s | 2 | $1s_{1/2}$ | 2 | | ← (2) |

| Reasonable nuclear | Reasonable nuclear plus spin-orbit | Accumulated occupancy | Magic numbers |

Odds, evens, and shells

4. **Shell model.** The order of energy levels, and the number of possible locations at each one, for a single particle in various situations. The results on the left are in a nuclear potential but where the effects of spin are ignored; while the results on the right are when the effects of spin and rotary (orbital) angular momentum are included. 'Magic numbers' are where noticeable gaps appear (such as above the 2, 8, 20).

making ten protons or ten neutrons in all. However, this shell also contains a duo with no angular momentum, so that in total twelve protons or neutrons can be accommodated. Adding this to the eight states already accounted for, and we find a total of twenty.

Here again we see this verified as the doubly magic isotope calcium-40, $_{20}^{40}Ca$, forms 97% of all naturally occurring calcium, which is the fifth most common element in the earth's crust.

The pattern of shells naively continues onwards and upwards in energy and numbers of constituents, but as the density of states becomes larger, the role of magic numbers becomes less marked. This is because the propensity for nucleons to attract one another depends not only on their location in shells, but also on their intrinsic rotation. Nucleons can spin (in units of Planck's quantum their spin is one-half) and quantum theory implies that this can be oriented in either of two senses—which we might idiomatically refer to as clockwise and anticlockwise. Neighbouring nucleons feel an enhanced or reduced attraction, depending on whether their spins are in the opposite or the same sense, and also on whether the quantum shell within which they belong is itself rotating and, if so, whether this is in the same or the opposite sense to the individual nucleons. To determine the total effect involves careful accountancy in the quantum ledgers. Suffice to say that for more than twenty protons or neutrons, the effect of these couplings begins to obscure the underlying shells, and a more empirical approach pays dividends (which we shall see later in this chapter).

Nonetheless, amid this melange of states some relatively sharp magic numbers survive, such as 82 and 126. These large values are manifested in the stability of lead, for which $_{82}^{208}Pb$ is doubly magic: 82 protons and 126 neutrons combining to the total of 208 constituents. Lead-208 is the heaviest stable isotope. The heavier bismuth-209 is often listed as stable, and for practical purposes it may be regarded as such, but it is actually unstable with a lifetime of the order of 10^{19} years, which is some billion times longer than the age of the universe.

Among combinations of more than 208 nucleons, uranium-238 survives in the earth's crust even though it is unstable. Its half-life is about 4.5 billion years, which is comparable to the age of the earth. Also, thorium-232, $^{232}_{90}Th$, which has a half-life of 14 billion years, slightly longer than the age of the universe, decays by emitting alpha particles. It produces the radioactive gas radon-220 among its decays. Uranium-238 and thorium-232 are so long-lived as to be effectively stable on the timescales of human affairs, but over the aeons their decays spawn a host of short-lived radioactive elements such as radium and radon (chapter 5) en route to the stability of lead.

The empirical values of magic numbers give guidance when building detailed theories of the nuclear force. In turn these concepts predict that there may exist further magic numbers, leading to an 'island of (relative) stability' for isotopes beyond uranium. The magic 126 of neutrons, as in lead, could occur for 126 protons, but only if a much larger number of neutrons is present to help stabilize the isotope. Theory suggests that 184 may be magic, in which case 'unbihexium'—for 'one-two-six-ium'—$^{184}_{126}Ubh$ —may have a relatively long half-life. We shall see more about this in chapter 6.

Semi-empirical mass formula

The mass of a nucleus is not simply the sum of the masses of its constituent nucleons. $E = mc^2$ links mass to energy, and some energy is taken up to bind the nucleus together. This 'binding energy' is the difference between the mass of the nucleus and its constituents. Thus if proton, neutron, and nucleus have masses $m_{p,n,A}$ respectively, the binding energy, B, is

$$B = (Zm_p + Nm_n - m_A)c^2$$

where Z, N are the number of protons and neutrons.

The larger the binding energy, the greater is the propensity for the nucleus to be stable. Its actual stability is often determined by the relative size of the binding energy of the nucleus to that of its near neighbours in the periodic table of elements, or of other isotopes of the original elemental nucleus. As nature seeks stability by minimizing energy, a nucleus will seek to lower the total mass, or equivalently, to increase the binding energy. This it does by emitting an alpha particle cluster, by beta decay, or in more extreme cases, by splitting in two, as in the fission of uranium. An effective guide to stability, and the pattern of radioactive decays, is given by the semi-empirical mass formula (SEMF).

Figure 5 shows the binding energy per nucleon as a function of the total number of nucleons, A. It reveals the distinct shell structure for light nuclei, where the binding energy peaks at four, twelve and twenty. Beyond this the individual shells become lost among a more uniform dependence on A. For $A \geq 20$, the binding energy per nucleon, B/A, is approximately independent of A,

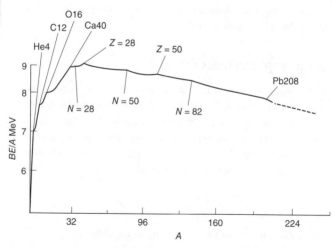

5. **Binding energy per nucleon for the most stable isotopes as a function of the mass number A. The broken line indicates that all nuclei heavier than lead are unstable.**

although there are clear corrections to this trend: B/A peaks around iron and gradually falls away for heavier nuclei.

That B/A empirically is roughly constant gives an immediate picture of the nucleus: nucleons are closely packed, and the interactions among them are short range. To see why, imagine you are a nucleon, surrounded by a crowd. You only interact with your nearest neighbour, and so you do not care how large the crowd is. Thus the amount of your interaction, in effect the binding energy per nucleon, is independent of A.

This picture is fine for a nucleon within the nucleus, but not for one on the edge of the crowd. There you have only a neighbour on the inside, and none beyond you. Thus your chance of interaction will be less (Figure 6). For a tightly packed nucleus we can estimate the dependence on A as follows.

For a sphere, the volume grows in proportion to the cube of the radius, R^3, while the surface area grows in proportion to R^2. Thus

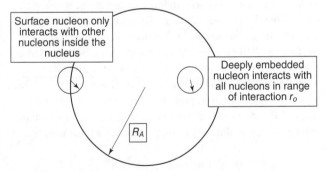

6. Nucleus of radius R_A containing nucleons with a range of interaction r_0. Central nucleons can interact with all other nucleons within the interaction radius. Nucleons at the surface can only interact with those within the nucleus. The number of surface nucleons is proportional to the surface area of the nucleus $4\pi R_A^2$.

if A nucleons tightly fill the volume, the fraction on the surface is in proportion to $A^{2/3}$. As these have reduced contributions to the binding energy we have to subtract them in the accounting, whereby the mass formula so far may be summarized as

$$B = aA - bA^{2/3}$$

where a, b are empirical constants with dimension of energy, where $a \sim 15\,\text{MeV}$ and $b \sim 17\,\text{MeV}$.

Cue protons

The most obvious omission, so far, is the distinction between neutrons and protons, specifically that the latter mutually repel. This electrostatic energy reduces the binding energy. Intuitively, having too many protons destabilizes the nucleus.

The electrical effect is a long-range force, which dies as the square of the distance between the charged particles. Force is the change of energy with distance, thus the electrostatic energy varies as $1/R$. For a densely packed nucleus, this is proportional to $1/A^{1/3}$. So we have to remove from the binding energy a quantity $dZ^2/A^{1/3}$, where d has dimensions of energy. Technically as you can only interact with all but yourself, this numerator should be $Z(Z-1)$ but for large Z this is a fine detail. This electrostatic contribution becomes more important for larger A, as A itself grows with Z. For light nuclei the binding energy grows with A until electrostatic repulsion takes over in large nuclei. Thus we see the qualitative shape of Figure 6 where the binding energy rises to a maximum, at intermediate A, due to this competition. This is summarized by

$$B = aA - bA^{2/3} - dZ^2 / A^{1/3}$$

The empirical magnitude of $d \sim 0.7\,\text{MeV}$ is consistent with that computed in a model where the nucleus is a uniform sphere of electric charge. This empirical success helps to support the picture that neutrons and protons are packed tightly into spherical lumps when in atomic nuclei.

At large values of Z, the penalty of electrostatic charge, which extends throughout the nucleus, requires further neutrons to add to the short range attraction in compensation. Eventually, for $Z > 82$, the amount of electrostatic repulsion is so large that nuclei cannot remain stable, even when they have large numbers of neutrons. The binding energies around $Z = 92$; $A = 238$ (uranium) and $Z = 90$; $A = 232$ (thorium) are finely balanced, which makes the instability of uranium and thorium sufficiently delicate that these elements remain in significant amounts on earth. All nuclei heavier than lead are radioactive.

To see how radioactivity is manifested, we need to include two effects of quantum mechanics in our accounting. This will complete the SEMF and expose the criteria for alpha and beta decays.

Quantum pairs

The existence of shells at large A is lost among the plethora of states. The physics that underpins them, however, shows up in the patterns of beta decay. The fundamental underlying principle is quantum *exclusion*: no two identical nucleons can occupy the same quantum state. To see how this affects the masses of isotopes, imagine once again the available energy levels for nucleons to be like the rungs on a ladder. One ladder is for protons, one for neutrons.

As one adds more neutrons or protons to the mix, to minimize the energy they must go onto the lowest unoccupied rung, namely the one with the lowest energy. It is easy to see that for a given number of nucleons, the minimum total energy will be when their numbers are the same, when neutrons and protons are added in symmetry. For example, suppose there is the same number of each and then we add two more. If one is a neutron and one a proton, this costs one more rung's-worth of energy apiece, a total of two. However, if both are neutrons, say, the first will cost one unit but the second will cost two, a total of three. The same will be true for two protons. By adding more and more, you can check that the growth in energy per nucleon is proportional to the square of the asymmetry:

$$\frac{(N-Z)^2}{N+Z} \equiv \frac{(A-2Z)^2}{A}$$

This effect competes with the electrostatic energy. The latter is proportional to Z^2, and such nuclei also require an excess of neutrons to stabilize them. Too large an excess, however, and the $(N-Z)^2$ asymmetry begins to add unwanted energy. Thus for some chosen number of nucleons—a fixed A—this destabilizing energy is large at very large Z, due to the electrostatic contribution, and is also large at very small Z, because of the huge asymmetry in numbers of protons and neutrons. The shape is illustrated in Figure 7.

Nature seeks to adjust the nucleons to a minimum energy. Thus isotopes with small Z move towards stability by increasing the value of Z, which is done by beta decay:

$$A(Z) \to A(Z+1) + e^- + \bar{\nu}.$$

Conversely, isotopes with large Z shed charge by positron emission:

$$A(Z) \to A(Z-1) + e^+ + \nu.$$

This continues in sequence across the elements until the state at the bottom of the valley is reached.

This is true for nuclei where A is an odd number. For nuclei where A is even, there is one final piece of accounting required: is the nucleus odd-odd or even-even?

Pairs of identical nucleons have an enhanced mutual attraction when their spins are oriented in opposite directions. Thus there is an extra binding for an even number of protons relative to an odd number, because for an odd number one lone proton will lack this attractive partner. The same holds true for neutrons. Thus in place of the single valley of Figure 7, we now have two, similar in shape but displaced in energy: the even-even has two additional attractions relative to the odd-odd situation.

As before, nuclei move towards the minimum by beta decays. With each decay a single proton is replaced by a neutron, or vice versa. Thus an odd-odd configuration becomes an even-even, and then

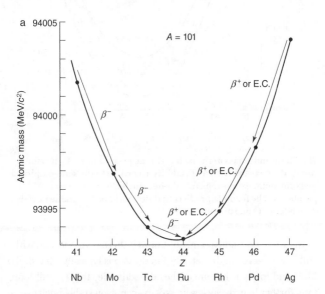

7a. **Pattern of beta decays and stability of odd and even nuclei,** $A = 100$ **or** 101**. (a) The atomic mass of the isobars of** $A = 101$ **for a range of** Z **near the region of stability according to the semi-empirical mass formula (the line drawn through the points is to guide the eye). Negative beta emission increases** Z**; positive beta emission or electron capture (E.C.) decreases** Z**. There is one stable isobar for the odd nucleus** $A = 101$**:** $Z = 44$**, ruthenium.**

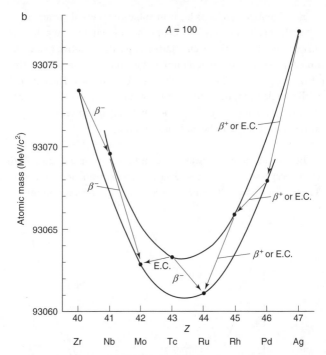

7b. The atomic mass of the isobars for an even value of $A = 100$ for a range of Z near the region of stability according to the semi-empirical mass formula. Even-even and odd-odd nuclei lie on different parabolas. The formula predicts that there will be two stable nuclei: molybdenum and ruthenium.

moves on to odd-odd again. Or it does if there is an odd-odd with still lower energy in the set. The energy gap is typically a handful of MeV, and so at the bottom of the odd-odd valley there is still room for a further beta decay leading to an even-even stable nucleus.

Thus we see why odd-odd nuclei are so scarce. Indeed, you might now wonder why there are any at all! Their five members are almost exclusively the isotopes of light elements, where the shell structure still dominates. Thus the deuteron ^2_1H is the

56

first, along with stable isotopes of the odd numbered elements $^{6}_{3}Li$; $^{10}_{5}B$; and $^{14}_{7}N$. The sole heavy example is $^{180}_{73}Ta$, whose stability is due to the large difference between the amount of its spin and that of its near neighbours.

Alpha decay

The binding energy per nucleon is biggest when $A \sim 60\text{--}100$. Thus heavier nuclei will tend to convert into these lighter more stable nuclei, which they can do by shedding neutrons and protons in the form of alpha particles.

The process of alpha emission is a paradigm of the interplay of quantum mechanics and classical ideas. Once the positively charged alpha particle has escaped from its strong prison, the repulsion of like charges ejects it violently from the residual nucleus, which is itself highly charged. This is the source of the energetic alpha particles that proved so useful as nuclear probes in the early years of the 20th century. But first the alpha has to escape the strong binding of its parent nucleons.

We might make analogy with climbing over the Alps, from the Chamonix valley into Italy. The alpha particle initially is trapped within the valley. Classical physics would imply that the alpha particle will remain where it is—trapped inside the heavy nucleus—unless enough energy is supplied to climb up and over the mountain peak to reach the downward slopes of the far side. However, quantum mechanics allows it to escape by a process known as *tunnelling*. It is as if it can exit via the Mont Blanc tunnel, but only if it does so in a time less than that constrained by quantum mechanics. This is another example of quantum uncertainty—where the alpha particle borrows an amount of energy, ΔE (which in the classical example would be the amount required to lift it up and over the mountains) for a short time Δt, the product constrained by the magnitude of Planck's constant: $\Delta E \times \Delta t \leq h / 4\pi$.

The net result of alpha decay is this. A heavy nucleus, X, containing A nucleons, N neutrons and Z protons decays to one containing $A - 4$, and with reduced charge:

$$_Z^A X(N) \rightarrow {}_{Z-2}^{A-4} X(N-2) + \alpha$$

Technetium

The element technetium is named after the Greek for 'artificial'. This name is slightly misleading, as the element is real enough, but is (almost) absent on earth as it is the lightest element that is totally radioactive, and thus without any stable isotopes. Thus any significant quantities of it have to be made in the laboratory, for example by bombarding other elements with neutrons. To this extent it is artificial rather than natural. As to why it is like this, the answer in part is because it has an odd number of protons, which implies that one proton will be unpaired, but obviously this is not the whole story as there are many stable 'odd' elements. However, the values of $Z = 43$ and $A = 97$ in the semi-empirical mass formula do imply that technetium-97 is very close to the line between stability and instability. In summary, the numbers of protons and neutrons in technetium happen to have no stable configuration; the behaviour of large numbers of protons and neutrons is complicated to describe, and the semi-empirical mass formula gives a good description. While we might not a priori have predicted that technetium is unstable, if we were looking for an example of an element that would theoretically be unstable, technetium would fit the conditions.

Dmitri Mendeleev in the 19th century had noted a gap in his periodic table and successfully predicted the properties of this missing element, a metal, which lies adjacent to molybdenum and below manganese. At the time, there seemed to be nothing unusual about it, and the continued failure to find it became almost a paradox. It was eventually discovered in 1936, in foils of molybdenum, which had become radioactive during their use in experiments at a cyclotron in Berkeley.

The key was that molybdenum had been irradiated with neutrons, which today is the route to producing technetium. Molybdenum is found in ores, and 24% of it is $^{98}_{42}Mo$. Neutron bombardment produces ^{99}Mo which is radioactive and beta decays with a half-life of about 66 hours into Technetium-99

$$n + {}^{98}_{42}Mo \rightarrow {}^{99}_{42}Mo \rightarrow {}^{99}_{43}Tc + e^- + \overline{\nu}$$

The ^{99}Tc itself undergoes beta decay, but with a lifetime of some 200,000 years. There is an excited state of this isotope, labelled ^{99m}Tc, with a half-life of six hours, which is used in medical diagnostics.

The minute quantities of technetium that are in the earth's crust consist of technetium-98. Its half-life of 4.2 million years may appear long to human timescales, but relative to the age of the earth this corresponds to about 1000 half-lives. So of technetium in the primaeval magma over four billion years ago, only one part in 2^{1000}—or 10^{325}—survives. Compared to such values, the number of atoms of all elements in the entire earth is trifling—'only' about 10^{50}. Thus we can confidently assert that no 'original' technetium exists here. The traces of technetium are produced by other natural processes, such as the spontaneous fission of uranium in ores, or by collisions between cosmic rays and atoms in the atmosphere, or molybdenum in the earth's crust. The dominant source of terrestrial technetium is probably radioactive waste from nuclear reactors.

In the universe at large, technetium can be produced in some stars, known as *technetium stars*.

Chapter 5
Making and breaking nuclei

Big bang nucleosynthesis

We have seen how the nuclear atom was discovered, how nuclei are glued together in defiance of electrostatic repulsion, and identified a still deeper level of reality—quarks and gluons that are the constituents of nucleons. In this chapter we will see how the smorgasbord of atomic elements came to be.

Most recently (by which I mean the last five billion years!) the majority of elements found on earth were formed inside a long dead star, where they were all cooked from protons. The protons were synthesized within the first second of the universe, and their constituent quarks, and also electrons, were made even earlier than that.

The heat energy in the big bang, some 13.7 billion years ago, converted into counterbalanced particles of matter and antimatter, courtesy of $E = mc^2$. The seeds of atomic nuclei were initially the simplest constituents: quarks. Somehow the symmetry between quarks and antiquarks was lost (at least in the observable universe today where matter dominates in bulk). We don't know how this happened. The story of how the quarks synthesized matter as we now know it is well understood, however.

Initially, the universe was so hot that the quarks and gluons were in a quark–gluon plasma. After about 10^{-2} seconds, the universe had cooled to a temperature of about a trillion degrees, at which these fundamental constituents coalesced into the 'frozen' nucleons, the seeds of nuclear matter.

The following processes took place:

$$e + p \rightleftharpoons n + \nu$$

where the double arrow illustrates that this process could occur in either direction.

As a neutron has slightly more mass than the combined masses of a proton and an electron, the 'natural' direction for the processes is from right to left: the neutron has a natural tendency to lower the mass of the whole, liberating energy via $E = mc^2$. However, the heat of the universe was such that the electrons and protons had considerable amounts of kinetic energy. Thus their total energy exceeded that locked into the mass (mc^2) of a neutron. Consequently, the process could as easily run from left to right (electron and proton converting into neutron and neutrino) as the other direction.

The universe continued to expand and cool, which made it harder for the production of neutrons to continue. After about a microsecond this neutron production reaction was effectively frozen out. The surviving reaction was

$$n \rightarrow p + e + \bar{\nu}$$

with a half-life of about ten minutes. However, within three minutes most neutrons had been captured by protons to form stable isotopes of light elements, such as helium, so during this brief epoch the neutron was effectively stable. All of the neutrinos were by now free, and became the first fossil relics of the universe.

About a billion neutrinos were produced for every atom that would eventually form.

After a few minutes, the universe had cooled enough that any neutrons and protons that collided and felt a strong attraction would be able to survive as a couple—the deuteron. Neutrons within a deuteron are stable. Thus, with the emergence of the deuteron, neutrons were safe, and the basic stable seeds were in place by which the nuclei of light elements could be synthesized.

At this stage, the universe at large played out a sequence similar to that which takes place in the Sun today. Hydrogen may be relatively uncommon on earth (except when trapped inside molecules such as water, H_2O) but in the universe at large it is the most common of all. In the big bang, and in the heat of the sun's core, hydrogen's proton forms the common seed from which all nuclei are ultimately formed. Any neutrons trapped initially within deuterons, and protons, build up nuclei of helium. In the early universe, this continued until all of the neutrons had either decayed or been trapped inside stable isotopes, or until the particles in the expanding universe were so far apart that they no longer interacted with one another.

I shall first describe these processes in the early universe, and then compare with stellar synthesis, where there are strong similarities but also differences of detail.

Synthesis of light elements

Once protons and deuterons are in the mix, collisions between them lead to ^3_2He. Two protons cannot form ^2_2He as it is too unstable, but two deuterons can combine. In this case we again have ^3_2He, but with the liberation of a free neutron, or alternatively a triton (seed of tritium) and a proton. These collisions have thereby led to isotopes with $A = 3$. Rather than using words, it is easier to keep track of the accounting by simply recording the numbers of nucleons and protons (to identify the element) as they basically combine and rearrange themselves in clusters. Thus, the story so far (writing the proton as ^1_1H and deuteron as ^2_1H):

$$^1_1H + {}^2_1H \rightarrow {}^3_2He + \gamma$$

(where the conservation of energy leads to emission of the photon, denoted γ), and

$$^2_1H + {}^2_1H \rightarrow {}^3_2He + {}^1_0n$$

or

$$^2_1H + {}^2_1H \rightarrow {}^3_1H + {}^1_1p.$$

Alternatively we can illustrate these by the rearrangement of light and dark discs (Figure 8).

Although tritium is unstable, its half-life of twelve years is vast compared to the brief epoch of three minutes while these primitive isotopes were being formed. Thus both helium-3 and tritium can collide with protons, deuterons, and one another to build small quantities of lithium, beryllium, and perhaps even trifling amounts of boron.

This next stage begins with the synthesis of $A = 4$.

Some ^4_2He exists already from the collisions of deuterons:

$$d + d \rightarrow {}^4_2He + \gamma.$$

The emergence of tritium and ^3_2He add to this through

$$t(^3_1H) + d(^2_1H) \rightarrow {}^4_2He + n$$

63

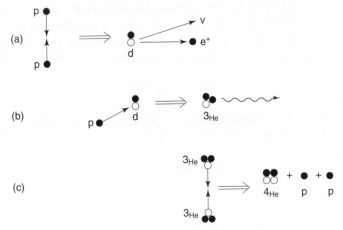

8. **Stellar nucleosynthesis.** The proton–proton chain reaction in the Sun starts when two protons, denoted p, fuse to make a deuteron (consisting of a neutron, the white circle, and a proton, the dark circle) together with a positron (e^+) and neutrino. Further steps leading to helium-4 are shown in (b) and (c).

and

$$\, {}^{3}_{2}He + d \rightarrow {}^{4}_{2}He + p.$$

The collision of two isotopes with $A = 3$ also makes ${}^{4}_{2}He$ as in, for example:

$$\, {}^{3}_{2}He + {}^{3}_{2}He \rightarrow {}^{4}_{2}He + p + p$$

The seeds of the next lightest elements, lithium and beryllium, also form. The amounts, however, are relatively sparse, as they are formed only after the lighter seedlings have emerged. Furthermore, these heavier isotopes can be disrupted by collisions with protons or neutrons, the fragmentation adding to the pool of helium-4. Thus, for example:

$$t({}^{3}_{1}H) + {}^{4}_{2}He \rightarrow {}^{7}_{3}Li + \gamma$$

produces lithium-7 but only so long as this avoids destruction by collision with a proton, as such a collision leads back again to helium-4:

$$_3^7Li + p \rightarrow {}_2^4He + {}_2^4He.$$

Beryllium is formed by analogy. In place of tritium, consider helium-3, in which case instead of lithium-7 we get beryllium-7:

$$_2^3He + {}_2^4He \rightarrow {}_4^7Be + \gamma$$

with a similar propensity for collisions with neutrons to convert into helium-4:

$$_4^7Be + n \rightarrow {}_2^4He + {}_2^4He.$$

Three minutes after the big bang, the material universe consisted primarily of the following: 75% protons; 24% helium nuclei; a small number of deuterons; traces of lithium, beryllium, and boron, and free electrons.

The relative abundance of helium to hydrogen depends on the expansion rate of the universe. The principles of physics that determine its expansion are in some ways similar to those that control the behaviour of a gas in a container. The rate depends on the pressure, which depends on the temperature in the gas and the number of neutrinos inside the gas volume (the density). This in turn depends on the number of lightweight neutrino species. The observed amounts fit with predictions if there are three varieties of neutrino. Experiments in particle physics show directly that there indeed just three varieties of light neutrinos, which confirms the cosmological description of nucleosynthesis in the big bang theory.

The abundance of deuterium depends on the density of 'ordinary' matter in the universe. (By ordinary we mean made of neutrons and protons as against other exotic things such as dark matter.)

The numbers fit with the empirical data provided that the density of ordinary matter is much less than the total in the universe. This is part of the dark matter puzzle: there is stuff out there that does not shine but is felt by its gravity tugging the stars and galaxies. It seems that much of this must consist of exotic matter whose identity is yet to be determined.

Then, 300,000 years later, the ambient temperature had fallen below 10,000 degrees, that is similar to or cooler than the outer regions of our sun today. At these energies the negatively charged electrons were at last able to be held fast by electrical attraction to the positively charged atomic nuclei whereby they combined to form neutral atoms. Electromagnetic radiation was set free and the universe became transparent as light could roam unhindered across space.

The big bang did not create the elements necessary for life, such as carbon, however. Carbon is the next lightest element after boron, but its synthesis presented an insuperable barrier in the very early universe.

The huge stability of alpha particles frustrates attempts to make carbon by collisions between any pair of lighter isotopes. A pathway in theory might appear to be the formation of beryllium-8 from a pair of alpha particles, followed by the absorption of a third alpha:

$$_2^4He + {}_2^4He(+{}_2^4He) \rightarrow {}_4^8Be(+{}_2^4He) \rightarrow {}_6^{12}C.$$

The problem, however, is that beryllium-8 falls apart into a pair of alphas almost as soon as it is formed. To make carbon requires three alphas to collide within a time less than the lifetime of beryllium-8; in effect, simultaneously.

There was in practice no chance of a triple coincidence of alphas fusing in the diffuse atmosphere after the big bang. Thus no carbon or heavier isotopes were formed during big bang nucleosynthesis. Their synthesis would require the emergence of stars. In the hot dense centre of a star, however, this barrier can be overcome. Three alpha particles may collide and fuse to make carbon-12, and with carbon in the mix, the way to synthesizing isotopes of heavier elements is open. This brings us to stellar nucleosynthesis.

Stellar nucleosynthesis

In the heat of the big bang, quarks and gluons swarmed independently in quark–gluon plasma. Inside the sun, relatively cool, they form protons but the temperature is nonetheless too high for atoms to survive. Thus inside the sun, electrons and protons swarm independently as electrical plasma. It is primarily protons that fuel the sun today.

Protons can bump into one another and initiate a set of nuclear processes that eventually converts four of them into helium-4, analogous to those we met in the big bang example.

The main difference is that tritium was effectively stable in the short timescales of big bang nucleosynthesis, and so it played a role there, whereas its instability limits its practical relevance over long timescales in the sun. Helium-3 is in practice more important than tritium in stellar nucleosynthesis.

As the energy mc^2 locked into a single helium-4 nucleus is less than that in the original four protons, the excess is released into the surroundings, some of it eventually providing warmth here on earth. The three major steps in reaching helium-4 are these:

$$p + p \rightarrow d + e^+ + \nu$$

$$p + d \rightarrow {}^3_2He + \gamma$$

$$ {}^3_2He + {}^3_2He \rightarrow {}^4_2He + p + p$$

The first step determines the timescales of solar burning. First, as like charges mutually repel, the two protons have to combat this electrostatic barrier before they can get close enough to initiate the reaction. The temperature of 10 million degrees gives the protons enough kinetic energy to do this. The total energy of the two protons, which includes that in the electrostatic field, exceeds that of a deuteron. As a result, one of the protons can turn into a neutron, by positron emission, which then binds to another proton, increasing the stability (as we saw in chapter 3).

This first part of the solar fusion cycle produces antimatter! The positron is almost immediately destroyed as it collides with an electron in the plasma, producing two photons which are scattered by other electrons and eventually work their way to the solar surface after several thousand years. By this time their energy is much reduced and what originated as photons with energy of about 500 keV have degraded to less than an eV and help form part of sunlight. That bouncing around has taken thousands of years; by contrast the neutrinos pour out from the centre unhindered and reach us within a few minutes. Detecting these neutrinos on earth, both in their numbers and energies, establishes experimentally that solar fusion proceeds by the steps outlined here.

The electrostatic barrier, and the weak force that controls the positron emission of the beta radioactivity, make the first stage of the proton–proton fusion processes relatively improbable. Five billion years after the sun was born, any individual proton has only a 50:50 chance of having taken part in this fusion. Put another way: this far the sun has used up half of its fuel. Once this first step is taken, however, events move on rapidly. The rearrangements of nucleons that build up heavier isotopes, such that a deuteron and a proton make ^3He and then lead on to form ^4He, happen almost instantaneously. It is the tardiness of the first step, $p + p \rightarrow d\nu e^+$ that controls the (slow) burning of the Sun that has been so important for allowing the long timescale of evolution to work its magic.

The Sun is shining courtesy of nuclear fusion. In another five billion years its hydrogen will have all gone, turned into helium. Some of the helium is already itself fusing with protons and other helium nuclei to build up the nuclear seeds of heavier elements, by steps analogous to those already described in big bang nucleosynthesis. The differences of detail, as we said earlier, are that because the sun produces these reactions continuously over aeons, unlike big bang nucleosynthesis, which lasted mere minutes, unstable isotopes, such as tritium, play no role in solar nucleosynthesis.

The synthesis of beryllium and boron involve beta radioactivity, and the neutrinos emitted in these processes have higher energy than those from the primary proton-to-helium phase.

So by detecting neutrinos from the Sun, and measuring their energy spectrum, we are able to get a quantitative look inside our nearest star.

These higher energy neutrinos arise from the production of beryllium-7, followed by its decay:

$$^3_2He + {}^4_2He \rightarrow {}^7_4Be + \gamma;$$

followed by

$$^7_4Be + e^- \rightarrow {}^7_3Li + \nu.$$

The lithium-7 then combines with a proton to form two nuclei of helium-4:

$$^7_3Li + p \rightarrow 2\,{}^4_2He.$$

In contrast to the fusion of two protons, whose electrical repulsion involves just one unit of positive charge apiece, for lithium, three units of charge are present. Thus the fusion barrier is higher, and hotter conditions are needed to provide enough energy to overcome it.

Thus this process occurs at temperatures in the range of 10 million to 23 million degrees. At higher temperatures the beryllium—with four positive charges—can fuse with a proton to make boron, which again undergoes beta decay, with the emission of a neutrino, and converts into yet more helium-4.

$$_4^7 Be + p \rightarrow \, _5^8 B + \gamma;$$

$$_5^8 B \rightarrow \, _4^8 Be + e^+ + \nu;$$

$$_4^8 Be \rightarrow 2 \, _2^4 He$$

Thus the formation of boron and beryllium (illustrated in Figure 9) leads to a collapse back to smaller nuclei, all feeding the production of helium in stars. The outcome is that eventually the star exhausts its protons, and consists primarily of helium.

With no protons left to maintain the fusion engine, the stellar core collapses and temperatures reach more than 100 million degrees. At this point something very important for the emergence of life occurs: the temperature and pressure are so high that beryllium-8 forms faster than it falls apart into pairs of helium-4 isotopes. There is now a chance for beryllium-8 to hit helium and form carbon:

$$_4^8 Be + \, _2^4 He \rightarrow \, _6^{12} C + \gamma.$$

At last the carbon barrier has been overcome. Nonetheless, this is difficult, and fusion is sparse, so there is still considerable time before there is enough carbon for it to burn and form yet heavier isotopes. Once this can take place, some carbon-12 fuses with helium to make oxygen-16, which is stable and releases energy:

$$_6^{12} C + \, _2^4 He \rightarrow \, _8^{16} O + \gamma$$

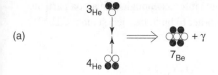

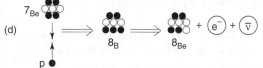

9. **Making helium via beryllium, boron, and neutrinos. The process in Figure 8 happens 85% of the time. Nearly all of the remaining 15% is due to helium-3 and helium-4 combining to make a single nucleus of beryllium-7 along with a photon (a). A series of reactions occurs until (e), the beryllium-8 splits into two nuclei of helium-4. Protons are denoted by solid dark circles; neutrons by open white circles.**

A difficulty with all these reactions is to overcome the barrier of electrical repulsion. This is harder for nuclei with larger values of electric charge, Z, and so these reactions involving carbon and oxygen only begin at temperatures exceeding 10^9 degrees. At such temperatures, up to a thousand times hotter than at the centre of the sun, carbon and oxygen interact enough with alpha particles of helium or with one another to build heavier elements. The extreme example is

$$^{16}_{8}O + ^{16}_{8}O \rightarrow ^{28}_{14}Si + ^{4}_{2}He.$$

The CNO cycle: nuclear reactions in stars

By the end of the 1930s the nucleus had been established as a dynamic structure, a repository of large amounts of energy, which could be liberated as the constituents rearranged to form more stable configurations. The possibility that the sun's energy could result from nuclear fusion was suggested, but the specifics were unclear. Today, we know that four protons burn to make helium-4, as we have already seen. This needs no catalyst. However, in 1939 Hans Bethe found another way for four protons to convert to helium, so long as there is some carbon-12 in the mix.

It is known as the CNO cycle, after the elements carbon, nitrogen, and oxygen, which cycle in a series of reactions, liberating energy and catalysing the conversion of protons to helium-4. Initially a candidate for stellar energy production, this is now realized to be an energy source only in stars that are much hotter than the sun. A cyclic series of reactions begins with $^{12}_{6}C$ picking up a proton to convert to nitrogen, whereby further protons and beta decays takes nitrogen to oxygen and then back to carbon-12.

The cycle starts with

$$^{12}_{6}C + p \rightarrow ^{13}_{7}N + \gamma.$$

Next the $^{13}_{7}N$ reaches the stable $^{14}_{7}N$ in two steps: a beta-positive decay to $^{13}_{6}C$ followed by fusion with another proton, so

$$^{13}_{6}C + p \rightarrow ^{14}_{7}N + \gamma.$$

Fusion with another proton moves this on to $^{15}_{8}O$, which e^+ decay takes to $^{15}_{7}N$. Absorbing a fourth and final proton brings the cycle to

$$p + {}^{15}_{7}N \rightarrow {}^{4}_{2}He + {}^{12}_{6}C.$$

This recovers the $^{12}_{6}C$, and along the way four protons have converted to a single helium-4. This is illustrated in Figure 10. The electrostatic coulomb barrier for a proton to invade nuclei with $Z = 6$ or 7 requires a much higher temperature than the 15 million degrees of the sun.

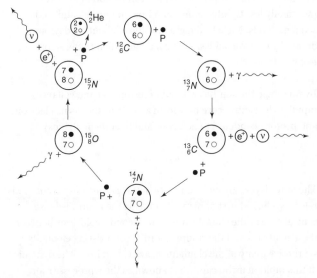

10. CNO (carbon, nitrogen, oxygen) cycle. Subscripts denote the number of protons; superscripts denote the total number of nucleons. The γ (gamma) denotes a photon and υ a neutrino. The wobbly lines illustrate the emission of energy as photons and neutrinos from the star.

Supernova nucleosynthesis

For stars that are about ten times more massive than the sun, after the sequence mentioned earlier has formed silicon, the star may collapse under its own weight. Following this gravitational collapse, the temperature can reach 3 billion degrees, which is enough for silicon to burn rapidly with helium. By sequential accretion of alpha particles, this can build elements all the way up to iron and nickel within a day. This sequence of isotopes is formed:

$$\,^{28}_{14}Si + \,^{4}_{2}He \rightarrow \,^{32}_{16}S \rightarrow \,^{36}_{18}Ar \rightarrow \,^{40}_{20}Ca \rightarrow \,^{44}_{22}Ti \rightarrow \,^{48}_{24}Cr \rightarrow \,^{52}_{26}Fe \rightarrow \,^{56}_{28}Ni$$

The next step in this process would lead to zinc-60, but this is marginally less tightly bound than iron and nickel, and so the sequence ends. Nickel-56 has a half-life of about six days and decays by positron emission to cobalt and then to the stable isotope: iron-56.

By this stage the star has run out of fusion fuel and collapses rapidly. The central core may form a neutron star, as its electrons and protons condense to neutrons and liberate neutrinos:

$$e^- + p \rightarrow n + \nu.$$

The outer layers are ejected in a *type-2 supernova* explosion. This releases a large burst of neutrons outside the neutron-dense central core of the star. Within a few seconds, collisions between these neutrons and the smorgasbord of pre-existing elements synthesize many of the elements heavier than iron. The dynamics of this rapid neutron capture is known as the r-process (r for rapid) and has been studied by experiments using radioactive beams in laboratories on earth. This is described in chapter 6. Large clusters of nucleons all the way up to the most massive metastable isotopes of uranium and thorium can be formed in such processes. They are ejected into the cosmos by the explosion, and once in space, they can be further modified by *cosmic spallation*.

Supernovae are not the only source of elements heavier than iron. About half of these heavy elements are made by the s-process, where neutron capture happens slowly (hence the s for slow). Slow here means relative to the timescale for beta decay. If a neutron is captured and the resulting isotope has time to undergo beta decay before another neutron hits, this will increase the electric charge on the nucleus and sequentially, over thousands of years, isotopes of heavy elements that are stable against beta decays can accumulate. This mechanism is important in red giant or *asymptotic giant branch* stars.

Cosmic spallation

Cosmic rays consist of high energy protons, alpha particles, some heavy nuclei, electrons, and photons. When they hit pre-existing matter, deep in space or hereabouts, fragments may be ejected from the struck isotopes; this is known as spallation. This chipping away at interstellar matter tends to reduce its atomic weight. Such collisions can lead to the formation of some light elements, primarily lithium, beryllium, and boron, adding to that produced during the big bang. Some tritium is also formed through spallation. These processes continue even now in interstellar space. This explains why interstellar space contains higher amounts of the light elements boron, beryllium, and lithium than are found in the atmospheres of stars, as these elements are not copiously produced in stellar nucleosynthesis.

In particular, spallation may liberate individual neutrons, protons or alphas which subsequently become attached to other clusters to form new isotopes. Spallation is the source of some radioactive isotopes in the atmosphere and in the earth's crust. One famous example is the formation of carbon-14, which is the key to radiocarbon dating of organic matter. This is how it comes about.

The impact of cosmic rays in the atmosphere may liberate neutrons. The air contains much nitrogen, and when struck by a neutron this produces carbon-14:

$$n + {}^{14}_{7}N \rightarrow {}^{14}_{6}C + p.$$

Carbon-12 is stable but carbon-14 has a half-life of 5730 years. Both of these isotopes combine with oxygen in the atmosphere to make carbon dioxide. Plants absorb this CO_2 while they are still alive, but after they die, there is no means of replenishing the carbon-14, which decays, while the carbon-12 stays intact. The ratio of ${}^{14}C$ to ${}^{12}C$ in the dead material thus falls, and by measuring the relative abundance of ${}^{14}C$ to ${}^{12}C$ and other elements in a sample, it is possible to determine the time elapsed since the last ingestion of carbon dioxide, in effect, its age.

Age of the earthly elements and the stars

Using nuclear radioactivity to measure the age of the earth is not nucleosynthesis, but an example of how isotopes, once synthesized, provide natural clocks by which their age can be determined. We have just seen one example with the decay of carbon-14 relative to carbon-12, by which the age of dead tissue can be determined. Now we turn to long-lived isotopes, trapped in rocks of the new-born earth, which have been slowly decaying ever since.

A common method is to measure the ratios of different stable isotopes of lead: lead-206, 207, and 208 or 204. These are the stable end products formed in the radioactive decays of uranium and thorium. In particular U-238 ends up as lead-206, with a half-life of 4.47 billion years, while U-235 with half-life of about 700 million years leads to lead-207. It is easy to see the idea qualitatively. Suppose that the amounts of uranium and lead were the same when the ancient rocks were formed. The radioactive uranium will die out, and the stable lead will correspondingly increase over time. Measurement of the ratio of U-238 and lead-207 today compared to that at the outset shows how long the clock for radioactive decay of uranium has been ticking.

This would be fine if it was indeed the case that the initial ratio of lead and uranium had been the same. In practice, this is unlikely, so we need to make a second comparison in order to get round this unknown. This involves a similar measurement of lead-206, which is the end-product of decays of U-235. The chemical processes that separate elements into different grains of the rock do not separate isotopes, so the initial amounts of different uranium isotopes would be the same, and those of lead would also have been the same. So by measuring the ratio of both lead isotopes the age can be determined.

An independent method is to use strontium and rubidium, two radioactive elements that are quite common in igneous rocks. Strontium has several isotopes including Sr_{87}, which is formed by radioactive decay of rubidium-87, and Sr_{86} which isn't. There are thus two sources of strontium-87 in rocks: primordial strontium, and that which has been produced by the radioactive decay of rubidium. The ratio of Sr_{87} and Sr_{86} was the same in all grains initially. So if we measure the ratio of Sr_{87} to Rb_{87} in the rock grain, we can determine how long the Rb_{87} has been decaying away, and hence determine the age of the rock, and ultimately the amount of time since the molten earth cooled.

These various methods show that the oldest rocks are about 3.8 billion years old. Similar measurements have been made for meteorites, which turn out to be as old as 4.6 billion years. This is essentially the age of the solar system.

We can even measure the ratios of various isotopes in stars by the 'autograph' of gamma rays that they emit. Different isotopes emit a spectrum of gamma rays of characteristic frequencies, or energies, such that they shine like beacons across the vastness of space and are detected here on earth. Once we know from this what the relative abundances of the isotopes of various elements are, we can perform the same sort of calculations as for the earthly rock samples and deduce how long the nuclear clock has been ticking in the star.

Formation on earth

Any isotopes of elements with half-lives less than 100 million years, and which originally accreted in the new-born earth, are today only found in trace amounts. Such isotopes are today made artificially, as in nuclear reactors, by the impacts of cosmic rays, or through the decays of some parent, as unstable isotopes seek to attain stability.

Supernova explosions can include very large isotopic clusters, in excess of 250 nucleons, which are highly unstable and rapidly fragment and decay to smaller lumps. There is an island of metastability at uranium and thorium, where half-lives are comparable to the age of the earth (U-238, 4.5 billion years) or even that of the universe (thorium-232, 14.5 billion years). The heaviest stable isotopes are those of lead: $^{204}_{82}Pb$; $^{206}_{82}Pb$; $^{207}_{82}Pb$; and $^{208}_{82}Pb$. Only the lightest of the lead isotopes is primordial, the others are the stable endpoints of decay chains of, respectively, U-238, U-235, and thorium-232. During these decay sequences, several isotopes are formed with short half-lives.

In all cases, the heavy isotopes lose atomic number by the emission of a sequence of alpha particles, meanwhile periodically lowering their total charge by beta decays. All of the isotopes in the intermediate chain are radioactive. They include elements with familiar names, such as radium, radon, and polonium, as well as the more exotic astatine, francium, and actinium. Francium and astatine, for example, have never been made in visible amounts as they would be vaporized by the energy from their own radioactivity. As we mentioned earlier, at any one time the total amount of francium in the earth's crust has been estimated to be less than 30 grammes, about one ounce, while that of astatine is even less, at about one gramme.

Much of the helium in our environment is the result of alpha radioactivity in the earth's crust. These positively charged nuclei of helium attract electrons to form helium gas. This explains the presence of helium in uranium ores, for example.

Radioactive decays can also cause fission to occur spontaneously. In such a case, a nucleus of uranium is split into two pieces, each of which contains a large number of nucleons (as, for example, in chapter 2). This is how samples of the radioactive elements promethium-61 and technetium-99 occur naturally on earth.

Many other isotopes are formed by human action. These include tritium, which resulted from thermonuclear explosions in the atmosphere half a century ago, and many other elements produced in nuclear reactors. By bombarding samples of uranium and other elements in a nuclear reactor, it is possible to synthesize isotopes with atomic numbers higher than uranium. These transuranic elements are all unstable, but naturally occurring traces may be present of neptunium, plutonium, americium, curium, berkelium, and californium. At least in theory, these can be present in uranium rocks, where neutrons, from cosmic rays or spontaneous fission, have interacted with the uranium to make these heavier isotopes. This is often stated as fact, although definitive evidence of having observed such traces is controversial. With these caveats in mind, however, these six transuranic elements should thus be added to the basic 92 in the list of elements that arise naturally on earth. This explains the total of 98, which may have been a surprise when you saw it mentioned at the start of this book. We shall describe the synthesis of transuranic elements, and of super-heavy elements in chapter 6.

Chapter 6
Beyond the periodic table

U-235, U-238, and plutonium

Uranium is the heaviest element that occurs naturally on earth in large quantities. The electrostatic repulsion acting throughout the pack of 92 protons is only just overcome by the localized strong forces, which makes these isotopes susceptible to nuclear fission (chapter 2). In U-238, which has an even number of neutrons, there is a stronger attraction relative to U-235, which contains an odd neutron. This difference is key to the different fission behaviour of these two isotopes and in their application to nuclear weapons.

A uranium nucleus is initially almost spherical. If energy is applied, the nucleus deforms into a dumbell shape with two ears. The electrostatic repulsion of the two mutual positive charges then pushes them apart until the strong force can no longer hold them. The nucleus splits into two large fragments—the phenomenon of fission.

For U-238 this energy is supplied by the impact of an initial 'fast' neutron. For U-235, however, it is sufficient to add a neutron with no energy other than its thermal energy—a 'slow' neutron. The reason, as already hinted, is that U-235, with its odd neutron, has relatively more energy than the ground state of U-236. So the addition of the neutron leads temporarily to an excited state of U-236. This may radiate a gamma ray to leave U-236 in its ground state, but it is more likely to fission.

The interest of fission is two-fold. First, the energy released is greater than that in other radioactivity. Second, the process also releases neutrons:

$$n + {}^{235}_{92}U \rightarrow {}^{144}_{56}Ba + {}^{89}_{36}Kr + 3n$$

which can hit other atoms of uranium and induce further fissions in a chain reaction. However, uranium found naturally is dominantly U-238; only about seven in every thousand atoms are U-235. So it is more likely that the secondary neutrons will hit U-238, and the chain reaction will end, rather than chance upon a relatively rare isotope of U-235, and maintain the chain. This is why the explosive release of energy, in an atomic bomb, required the uranium to be enriched—the fraction of U-235 had to be increased. This is a major industrial enterprise, which was one of the big challenges in the Manhattan project in World War II, and led to a search for alternative ways of making nuclear explosions. The first atomic bomb, at Hiroshima, exploited enriched U-235, whereas that at Nagasaki used plutonium.

The key to the use of U-235 was that the even number of protons was accompanied by an odd number of neutrons. Theorists conjectured, and experiment confirmed, that plutonium-239, again consisting of an even number of protons, 94, and an odd number of neutrons, 145, can also fission explosively by a chain reaction. Plutonium is synthesized, or *bred*, in a reactor, which contains U-238 as fuel. The U-238 may not fission, but it converts into plutonium by a sequence of two beta decays:

$$n + {}^{238}_{92}U \rightarrow {}^{239}_{92}U \rightarrow {}^{239}_{93}Np + e^- + \bar{\nu} \rightarrow {}^{239}_{94}Pu + e^- + \bar{\nu}$$

where $\bar{\nu}$ denotes an antineutrino.

The breeding of plutonium this way in a nuclear reactor is an example of the synthesis of transuranic elements, which today is a major area of research and of technical applications.

Transuranic elements from 93 to 100

Although uranium is the heaviest element that is found copiously, there are trace amounts of the next six elements, neptunium (Np), number 93, to californium (Cf) at 98. Einsteinium (Es) at 99 and fermium (Fm) at 100 were formed during hydrogen bomb tests half a century ago and were present in trifling amounts for a while afterwards, but do not occur naturally.

These elements are all synthesized in nuclear reactors or accelerators. Neptunium to californium are made when neutrons are captured by uranium, followed by beta decay, as in the production of plutonium as already mentioned. There is so much uranium on earth that these reactions can also be induced naturally by collisions of cosmic rays or by neutrons emitted in spontaneous fission. Such events can seed the production of transuranic elements all the way up to californium-98. Thus what a reactor synthesizes in macroscopic amounts by means of uranium targets and intense beams of neutrons, nature can do occasionally with uranium ores. Hence these six transuranic elements are included in the list of elements that occur naturally, making 98 in total.

The word 'naturally' here means occurring on earth. Supernovae produce all of these and more, as can other extreme events. For example, einsteinium-99 and fermium-100 were first synthesized in thermonuclear explosions. This discovery also gave support for the r-process in supernovae nucleosynthesis (chapter 5).

In the original surveys following the 'Ivy Mike' H-bomb test of November 1952, some evidence for a massive isotope of plutonium was found: $^{244}_{94}Pu$. This implied that six neutrons must have been absorbed rapidly:

$$^{238}_{92}U + 6n \rightarrow ^{244}_{94}Pu + 2e^- + 2\bar{v}.$$

This discovery was a surprise, and hinted that the intensity of a thermonuclear blast might enable more than six neutrons to be absorbed, and that with subsequent beta decays, elements beyond californium might be synthesized. This was duly confirmed when large amounts of radioactive material were found in coral.

Two novel elements were found to have been synthesized in the explosion. First we have einsteinium-99, which arose because fifteen neutrons had been absorbed, and seven beta decays had then occurred to improve the stability of the resulting heavy isotopes.

$$^{238}_{92}U + 15n \rightarrow ^{253}_{99}Es + 7e^- + 7\bar{v}.$$

The 15 neutrons had raised the mass number without changing the value of Z. This had led to an isotope that is overweight in neutrons. Beta decays then move through the elements, reducing the neutron excess until the optimum number of protons is achieved (chapter 4).

Second, the data also showed that in some instances as many as seventeen neutrons had been absorbed. This led to a heavier isotope: einsteinium-255. The extra two neutrons relative to einsteinium-253 makes this unstable to beta decay, whereby fermium-100 results:

$$^{255}_{99}Es \rightarrow ^{255}_{100}Fm + e^- + \bar{v}.$$

Fermium is the element with the highest Z that is made by neutron bombardment of uranium, and hence is the heaviest synthesized in macroscopic amounts. This limit arises because fermium decays by alpha emission, not by beta decay, and so it does not lead to the element with $Z = 101$: mendelevium. Thus

$$^{255}_{100} Fm \rightarrow {}^{4}_{2} He + {}^{251}_{98} Cf$$

and so does not produce mendelevium:

$$^{255}_{100} Fm \nrightarrow {}^{255}_{101} Md + e^{-} + \bar{\nu}.$$

In 1954, einsteinium and fermium were synthesized at Berkeley and Argonne in nuclear reactions which fused ions of nitrogen-14 with a target of U-238. They were also made by irradiation of plutonium or californium by intense beams of neutrons. Independently, in the same year, a Swedish team made fermium by bombarding U-238 with ions of oxygen-16. At this stage the results from the H-bomb tests, from 1952, remained classified secret, and the 1954 American papers coyly remarked that their 'discovery' was not the first study of the elements.

The synthesis of einsteinium and fermium from uranium on earth requires the capture of multiple neutrons. This is extremely unlikely, which is why neither of them is found in nature. They were the first elements made in high-power nuclear reactors or weapon tests. Nuclear explosions are the most intense man-made neutron sources, giving up to 10^{23} neutrons per cm^{2} per microsecond. This is 15 orders of magnitude greater than in high flux reactors. Such a reactor at Oak Ridge National Laboratory in Tennessee bombards targets of stable actinide elements with the aim of producing elements with $Z > 96$. (Actinide elements are the 15 metallic elements with atomic numbers between 83, actinium, and 103, lawrencium.) In these reactions, targets containing tens of grammes of curium are bombarded and produce typically 1/10 gm of californium, milligrams of berkelium ($Z = 97$), and picograms of fermium ($Z = 100$).

These are trifling amounts relative to what a 100 kiloton nuclear explosion can produce, where milligrams of fermium are synthesized. In high-flux reactors, micrograms of fermium have been made for use in specific experiments. Even such small amounts are sufficient for use as targets where bombardment with alpha particles or other ions enables trace amounts of elements with $Z > 100$ to be synthesized. These are super-heavy elements, all of which are synthesized at accelerators.

From transuranium to super-heavy

Elements number 95 and 96, americium and curium, are also produced copiously in nuclear reactors where uranium and plutonium are bombarded by neutrons. Each tonne of spent nuclear fuel produces about 100 grammes of americium and 20 grammes of curium. They are radioactive but with half-lives that range from months to many years, and so it is possible to accumulate macroscopic amounts, which can be used in industry or research. In particular they can used as targets for bombardment by heavy ions at accelerators. Thus for example, berkelium, element 97, was first made by bombarding americium with alpha particles. Beams of alpha particles, carbon-12, nitrogen-15, oxygen-18, neon-22, and magnesium-26 at laboratories such as Berkeley, Dubna in Russia, and GSI Darmstadt in Germany, have been fired at americium and curium and formed heavier elements. These super-heavy elements are described in the next section.

Americium decays to neptunium by the emission of alpha particles. It is such a powerful emitter of alpha particles that it can be mixed with beryllium to make a source of neutrons.

The key feature is that when an alpha particle hits beryllium, a neutron is released:

$$^{9}_{4}Be + {^{4}_{2}He} \rightarrow {^{12}_{6}C} + {^{1}_{0}n}.$$

It's possible that you have some americium in your home:

americium as a source of alpha particles is key to its use in smoke detectors. The alpha particle radiation passes though an ionization chamber—an air-filled space between two electrodes—which permits a small constant electric current to flow. Alpha particles are easily absorbed, so any smoke particles absorb them, which reduces the ionization and alters the current. This triggers the alarm.

Curium is one of the most radioactive elements. It has been used as a source of alpha particles in scientific instruments on space probes.

The transfermium wars

Traditionally the right to name a new element is that of its discoverer, or synthesizer. Elements 104 to 106 were found independently in the 1960s at Berkeley in the USA, and at Dubna in what was then the USSR. There was dispute about the robustness of some of the evidence and who had first established the existence of these elements. This led to a dispute over priority, and hence of names, as the two laboratories made different choices.

Element 104 was named rutherfordium in the USA, and kurchatovium in the USSR, after Rutherford and Kurchatov (father of the Soviet Union's atomic bomb). The two teams chose hahnium (USA) and bohrium (USSR) for element 105. Element 106, discovered at Berkeley, was named seaborgium after Glenn Seaborg, the nuclear physicist who had led the synthesis of several transuranic elements, including the first, neptunium and plutonium. This led to an independent controversy because Seaborg was still alive, and to name an element after a living scientist was regarded by some as inappropriate.

IUPAC (International Union of Pure and Applied Chemistry) attempted a solution. It proposed that 104 be named dubnium, after Dubna, the laboratory in the USSR where the synthesis had taken place, assigned rutherfordium to 106, and for 108 they chose hahnium, after Otto Hahn, discoverer of nuclear fission.

Meanwhile, elements 107 to 109 had been undisputedly synthesized at GSI Darmstadt, who had chosen bohrium for 107. For 108 and 109 they chose hassium, after an area of Germany, and meitnerium, after Lisa Meitner, colleague of Hahn and pioneer of nuclear fission. They objected that their name for element 108 was ignored as it was their undisputed discovery. The American Chemical Society also complained: element 106 was their unique discovery, for which they had chosen seaborgium.

After much debate, the right for a unique discoverer to choose the name was preserved. Thus 106 is seaborgium, and 107 to 109 recognize the original German choices. The two disputed claims, for elements 104 and 105, were diplomatically shared. The resulting names are rutherfordium for 104, the American choice, and dubnium for 105.

Island of stability

Lead-208, with 82 protons and 126 neutrons, is doubly magic, which aids its stability. Element 126, with a magic number of protons, thus offers the possibility of being relatively stable, although there will need to be an excess of neutrons to compensate for the electrostatic disruption. Thus the actual levels of stability and the isotopic contents are subject to much theoretical argument. The strategic consensus is that elements in the vicinity of 126 may show an island of relative stability. At the time of writing (2015) the heaviest element synthesized with certainty is number 117, 'un-un-septium'.

This nucleus was first produced in 2010, and was confirmed in 2014 at GSI, Darmstadt. The delicate experiment was actually an international collaboration. First it was necessary to prepare the target—of berkelium-249. This has a half-life of 330 days, and 13 milligrams were synthesized at Oak Ridge, Tennessee, and then sent to Mainz University. At Mainz, the berkelium was made into a target, which would be capable of withstanding bombardment by beams of calcium-48 at GSI.

$$_{20}^{48} Ca + {}_{97}^{249} Bk \rightarrow {}_{117}^{294} X$$

That a few examples of un-un-septium had been formed was confirmed by their radioactive decays—a sequence of seven alpha particle emissions leading to lawrencium was announced in May 2014 in Physical Review Letters, vol 112, page 172501. The sequence is

$$_{117}^{294} X \rightarrow {}_{115}^{290} X \rightarrow {}_{113}^{286} X \rightarrow {}_{111}^{282} Rg \rightarrow {}_{109}^{278} Mt \rightarrow {}_{107}^{274} Bh \rightarrow {}_{105}^{270} Db \rightarrow {}_{103}^{266} Lr.$$

To reach beyond un-un-septium en route to element 126 will require development of beams with projectiles heavier than calcium-48.

Neutron stars

It is possible that elements heavier than all of these exist in the cosmos in the form of so-called *neutron stars*. Although their name suggests that such stars are made solely of neutrons, this is not necessarily the case.

In our discussion of the forces that build atomic nuclei we have focused on the strong force, which attracts protons and neutrons in close contact, and the electrostatic repulsion among protons, which disrupts the collection. The weak force causes beta radioactivity. We have ignored the effects of gravity as this force is trifling between individual particles. However, the effects of gravity add up, such that for a large collection of particles, such as make a planet for example, it can dominate. Thus if more than 10^{57} neutrons are in a cluster, their cumulative gravitational

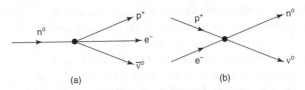

11. **Neutron stars and neutrinos are made together. When a neutron star is made, the beta decay process is at work. Electrons and protons in a dense star are squeezed so tightly together that they turn into a neutron and a neutrino. The neutrons form the neutron star; the neutrinos are radiated into space.**

attraction compares in overall strength with the strong force. The result is a macroscopic sphere, more than a kilometre across, comprised of neutrons—the neutron star.

As shown in Figure 11, the star is the end result of gravitational collapse whereby the plasma of electrons and protons in a star combines to form neutrons and a burst of neutrinos, as in chapter 5:

$$e^- + p \rightarrow n + v$$

Neutrons have no electrical charge, but are built of quarks, which do have charge. Although the electric charges of the quarks combine to zero, their individual magnetism does not. The result is that a neutron has a magnetic moment. The resulting magnetism of a rotating neutron star gives rise to the emission of electromagnetic radiation in the form of pulses—like a lighthouse beam that is only seen as the emitter turns in your direction momentarily. It is this phenomenon that led to the discovery of pulsars, later understood to be neutron stars. So much for astrophysics; we see here a remarkable synthesis with nuclear physics, where the semi-empirical mass formula implies that nuclei of 10^{57} neutrons can exist.

The collapse is unlikely to be so perfectly balanced that each and every proton finds an electron, so in reality there will be protons present as well. On the scale of 10^{57} even a billion is negligibly small. The term *neutron* star is perhaps a misnomer, as these objects probably contain protons and electrons too, with beta decays occurring, all leading to some state of equilibrium. So we can safely assert that there are elemental nuclei of elements far beyond un-un-septium, where stability is ensured by gravity and large numbers of protons are distributed within a kilometre-sized bunch of neutrons. For all practical purposes, however, these are collectively known as neutron stars.

Edge of stability: the drip line and the r-process

The landscape of isotopes can be represented on a figure, where the number of protons is on the y-axis and the number of neutrons along the x-axis. Each nucleus is represented by a box. A box is coloured black if the nucleus is stable, a dark colour for slight instability, and lighter colours for highly unstable. This immediately shows the tendency for the numbers of neutrons and protons to be similar, with a slight to moderate excess of neutrons. (An interactive chart, with links to data on the nuclei, is available at <http://www.nndc.bnl.gov/chart/>)

By adding more protons, or more neutrons, to a given stable nucleus, there will come a point where the nucleus formed by addition of one more such nucleon will immediately decay. The nucleon has in effect leaked or 'dripped' from the nucleus. These boundaries in the plot are referred to as the *drip lines* for protons or neutrons respectively.

The term drip line recalls the similarity of nuclei to liquid drops. Water molecules powerfully attract one another within the drop, but those on the surface only have neighbours within, and so feel less attraction. This gives rise to surface tension, where a drop may become more stable by splitting to form two smaller drops. Pursuing this analogy further, the nucleon drip line occurs when a nucleon drips out of the unstable nucleus, akin to water dripping from a leaky tap.

Neutron and proton drip lines play important roles in the formation of nuclei in stars and in supernovae explosions.

As we saw in chapter 5, the nuclei of elements heavier than iron are not formed by fusion of lighter nuclei because electrostatic repulsion is too great. For these nuclei to encroach one another, their kinetic energies would have to be large and the result would be a shattering of the nuclei rather than a melding into a single larger entity. Instead, the heavier elements are made by successive neutron captures, which make large neutron-rich unstable isotopes, followed by beta decays.

The formation of neutron-rich clusters is not inhibited by electrostatic repulsion. However, in the $[Z, A]$ landscape of isotopes, these configurations are far from the valley of stability.

Empirically, neutrons can be captured very rapidly in the core of a supernova collapse. This builds up isotopes along the neutron drip line faster than they undergo beta decay, and, as we have seen, is known as the r-process with r for rapid.

In supernovae, the r-process ends when nuclei are so big that they become unstable to spontaneous fission. This is when clusters of up to 270 constituents have formed. Once the flux of neutrons falls—for example, as the density of the supernova decreases—the highly unstable nuclei undergo a series of beta decays until they reach uranium and other relatively stable neutron-rich isotopes of heavy elements.

This qualitative theory of the formation of heavy elements can be studied in the laboratory by means of beams with heavy neutron-rich nuclei, such as U-238 for example, hitting a lighter target. The massive projectile chips off some constituents of the target, which recoil, while others transfer to the beam to form an exotic unstable heavy isotope. A simple example would be a deuteron target hit by a heavy ion. There is a chance that the proton in the target will recoil, and the neutron transfer to the beam, thus forming an isotope with one more neutron. By such experiments it is possible to determine the possible paths of r-processes through the landscape of neutron-rich, unstable nuclei.

This path depends on the amount of energy needed for a particular nucleus to capture a neutron, and its half-life. This in turn depends on the nuclear shell structure. The result is a zig-zag path through the landscape of A and Z. To understand the synthesis of heavy elements, it is necessary to understand how the structure of nuclear shells evolves far from stability. From the study of stable nuclei, it is known that 50 and 82 are magic numbers for protons. By symmetry of the strong force between protons and neutrons, these values should be magic for neutrons also. Thus neutron-rich nuclei such as $^{80}_{30}Zn_{50}$ and $^{130}_{48}Cd_{82}$ contain magic numbers of neutrons (we have made the numbers of neutrons explicit in the second subscript.) The r-process also creates neutron-rich isotopes that are doubly magic, such as $^{78}_{28}Ni_{50}$ and $^{132}_{40}Sn_{82}$. A current strategy in experimental nuclear astrophysics, is to use beams of radioactive heavy ions with the goal of studying how r-processes populate the landscape of highly unstable, neutron-rich, isotopes en route to the formation of stable heavy elements.

Chapter 7
Exotic nuclei

Halo nuclei

The liquid drop model of nuclei (chapter 4) assumes that a nucleus is a sphere of constant density. This implies that the radius of a nucleus consisting of A nucleons is

$$r = r_0 A^{1/3}$$

where $r_0 \sim 1.2\,\text{fm}$ (femtometres). While this is a good description of nuclei that are either stable or not far from the optimal conditions, where the number of neutrons exceeds the protons by up to about 50%, it does not describe some unstable isotopes with a larger excess of neutrons, or an excess of protons, such as those at or near the neutron or proton drip lines. In such cases with an overabundance of one variety of nucleon, halos may form around a central core.

An example is $^6_2 He$, which contains two protons and four neutrons. Two of the neutrons cluster with the protons in a compact alpha particle core, while the two remaining neutrons are loosely bound and encircle remotely. The lightest example of an isotope with a proton halo is boron-8. Four protons and three neutrons form a core, with the remaining proton in the halo. Nuclei with proton halos are more rare than their neutron cousins. This is due to the electrostatic repulsion among the excess protons, which destabilizes the halo.

All halo nuclei have short half-lives, typically of the order of milliseconds. They exist at the limits of stability in the collection of nuclear isotopes.

Borromean nuclei

Borromean rings are three circles where no two of them are linked but the trio is (Figure 12). Thus remove any one of the rings and the remaining pair will be unlinked, as illustrated in the figure. A Borromean nucleus by analogy is one consisting of three separate parts, bound in such a way that if any one is removed, the remainder become unbound. Carbon-12, which consists of three subunits of alpha particles, is an example. Remove one of them and the remainder is beryllium-8, which is not bound. Halo nuclei form Borromean structures. Their basic dynamics consists of two subsystems, one containing the central core, and the other a halo encircling the compact centre. In isolation the centre may be stable, but the halo not—as in $^{6}_{2}He$ with its alpha particle core and two satellite neutrons. It is possible that none of the subsystems can bind in isolation, even though the whole—cluster and halo—can survive, albeit briefly.

12. Borromean rings.

Studying these exotic configurations gives information about how many-body systems self-organize. While this has interest for the structure of atomic nuclei, it potentially has insights for the study of many-body systems throughout physics and chemistry.

Hypernuclei

If a quark in a neutron or proton is replaced by a strange quark, the resulting particle carries the property of strangeness, and is generically known as a *hyperon*. Within a nucleus, replace a nucleon by a hyperon and one has a hypernucleus.

The lightest example of a hyperon is the Lambda, denoted Λ, built from three quarks: one strange, one up, and one down. Most hypernuclei in practice tend to contain one or more Lambdas. Hypernuclei extend the chart of nuclear isotopes into a third dimension, which corresponds to the number of hyperons in the cluster. There is a theoretical possibility that strange matter, which contains a large number of hyperons, could be relatively stable. This is currently still conjecture, and not universally agreed among theorists. There is no evidence for strange clusters in cosmic rays, nor have examples been convincingly demonstrated in laboratory experiments, but the possibility that neutron stars might contain large clusters of strange matter is an open question.

In free space, a lambda undergoes beta decay and converts to a proton:

$$\Lambda \rightarrow p + e^- + \bar{\nu}$$

Within a nucleus this decay can be inhibited by the Pauli exclusion principle, analogous to the reason that a neutron can be stabilized within the nucleus even though it undergoes beta decay when it is in isolation.

Lambdas are held in nuclei by the strong force. Thus one interest in the study of hypernuclei is to deduce how the strong force between Lambdas compares to that for nucleons. As a Lambda has about 150 MeV more energy than a nucleon, locked into its mc^2, its beta decay within a hypernucleus releases a significant amount of energy, whereby the resulting conventional nucleus can be in a highly excited unusual state. Thus hypernuclei are interesting in their own right but also as a novel route into extending our understanding of more conventional nuclei.

Strange matter

Conventional nuclei contain large numbers of up and down quarks confined into triplets: neutrons and protons. The presence of a few strange quarks, also within triplets along with up and down quarks—such as a Lambda—form hypernuclei, as we have seen, which have been known and studied for decades. There is an unresolved question as to whether strange quarks can also exist in nuclei in clusters other than conventional hyperons. For example, the Pauli exclusion principle allows up to two identical flavors to co-exist in the lowest energy state, whereby a stable cluster of six quarks—two up, two down, and two strange—is in principle possible.

The Lambda is unstable and loses its strangeness, for example by beta decay, and converts to states built of up and down quarks. It is theoretically possible that matter with a large number of strange quarks doesn't behave in this way. When large numbers of strange quarks are present, the lowest energy state could contain nearly equal numbers of up, down, and strange quarks. The simplest example of such a state is known as a *strangelet*. Most theoretical models imply that, once formed, a strangelet changes to ordinary matter within a billionth of a second. There have been some sensational scare stories that the interactions between strangelets and ordinary matter cause the latter to change into strange matter, and that this is the most stable form of hadronic matter.

Even if this hypothesis were true, the half-life for this is much greater than the age of the universe. The reason for this remarkable stability becomes apparent when we see what the process involves. To get from a state of no strange quarks to a democratic mix of up, down, and strange requires first one, then two strange quarks, and so on, to be formed to seed the process. These initial strange quarks will form hyperons, such as a Lambda, which converts the nucleus to a hypernucleus. Hypernuclei are relatively heavy, however, and so for the transition to strange matter to happen, a large number of such conversions must happen simultaneously—faster than the timescale of beta decay.

This is extremely unlikely, at least under normal conditions and possibly in general. The requirements are similar to those of the r-process in the creation of neutron-rich nuclei for normal matter. To rapidly make matter rich in strange quarks thus requires unusual conditions.

One possible environment is the early universe, where strangelets could have been formed along with neutrons and protons. Another is in high energy collisions, such as at RHIC—the Relativistic Heavy Ion Collider in Brookhaven, New York—or the Large Hadron Collider at CERN, when beams of heavy ions are used, or in collisions between cosmic rays and the earth's environment. A signature would be the unusual behaviour of a strangelet in a magnetic field. Conventional nuclei are positively charged, and the ratio of their charge to mass determines how easily their trajectories curve in a magnetic field: the higher the charge for a given mass, the sharper the curve. Strangelets, however, have relatively small electric charges, and in the extreme of an equal number of up, down, and strange quarks, the total charge would be zero. Thus in general the ratio of charge to mass is very small compared to that of ordinary matter. This will lead to almost straight tracks in a magnetic field, which would be very distinctive. While the theoretical possibility of such strange matter has been recognized and searched after for over thirty years, no example has been found.

It is possible that within a neutron star, the pressure from the star's own gravity may cause the neutrons to collapse and form a coagulate of quarks. Whereas quark–gluon plasma forms at high temperatures and pressures, the quark star is a hypothetical state of cold matter. There is much theoretical debate as to the stability of such matter made of up and down quarks alone. Some models imply that the quark star is more stable if considerable numbers of strange quarks are also present, forming strange matter.

This touches on a scenario much loved by science fiction, where matter as we know it is not the most stable form and hence there is the possibility of some catastrophic collapse of the material universe into strange matter. In such a case, a collision between a strangelet and a nucleus on earth could cause the entire planet to become a lump of strange matter. Sadly for science fiction, but fortunately for us, all the evidence runs counter to this. The theoretical models that led to the concept of strangelets in the first place do not encourage this situation, and indirectly there are experimental counter indications. For example, were this the case, then neutron stars would all be strange stars, and in turn there should be a considerable flux of strangelets in cosmic rays. While absence of evidence can never be evidence of absence, all indications are that stable strangelets do not exist.

Antimatter nuclei

A fundamental property of quantum theory is that to every variety of particle there is a corresponding antiparticle—equal in mass, size, and shape but with the opposite sign of electric charge. The antiparticle of an electron is the positron, its positively charged sibling, which we met in positron emission and which is integral to positron emission tomography. This positron did not pre-exist within the nucleus but was created from the energy released in beta radioactivity.

The antiproton has the same mass as a proton but negative instead of positive charge. The antineutron likewise has the same mass as a neutron, and zero charge, which inspires a question: what distinguishes it from a neutron? Although the neutron has no electric charge overall, it has a small but measurable size within which are swirling electric charges, positives and negatives, carried by its quarks, which add up to zero. These give rise to magnetism, and as their motions are all reversed in an antineutron relative to a neutron, the antineutron and neutron can be distinguished by their different magnetic polarities.

The strong force acts on antinucleons with the same strength and effects as it does on nucleons. Thus, in theory, antimatter built of anti-atoms, where positrons encircle antinuclei made of antiprotons and antineutrons was as likely to have emerged after the big bang as the dominant material universe of matter currently known to us. Although individual antiparticles are regularly produced from the energy in collisions between cosmic rays, or in accelerator laboratories such as at CERN, there is no evidence for antimatter in bulk in the universe at large. In theory, however, a periodic table made of anti-elements should have the same bulk properties as those of the elements. One of the great mysteries in physics is how the symmetry between matter and antimatter was disturbed.

The simplest antinucleus is well-known: the existence of the antiproton was confirmed in 1955, and it has been used in experimental particle physics for decades. In experiments at CERN, the antiproton has been used to make an exotic atom of helium, in which one of the two electrons encircling the central doubly charged nucleus is replaced by an antiproton. This is achieved by first producing antiprotons, slowing them to low energy and then mixing them with ordinary helium gas. As an antiproton is negatively charged, it is attracted by the positively charged nuclei of helium atoms, much as electrons are. Momentarily the system is like an atom, where the antiproton orbits the central nucleus remotely. However, a massive antiproton orbits much closer to the nucleus than would an electron, and is rapidly entrapped by the strong nuclear force.

Thus one starts with an exotic form of atom and records X-rays emitted as the antiproton spirals inwards towards the nucleus. This lasts a few microseconds. Then briefly the atom consists of a single electron encircling an exotic nucleus of helium-4 and an antiproton. Within a picosecond, the antiproton is annihilated by one of the protons or neutrons, emitting gamma rays or pions.

This sequence of events can be used to learn much about fundamental physical properties. The spectrum of X-rays can be decoded to compute the mass of the antiproton. This is found to be the same as that of a proton to a precision of better than one part in a billion.

Anti-atoms of the simplest element—anti-hydrogen—have been made, and the difficulty in doing so illustrates some of the problems of creating more complicated antinuclei.

It is possible that, before 1995, not even one atom of antimatter had ever existed in the history of the universe. When positrons or antiprotons, which have been formed from the energetic collisions of cosmic rays, encounter one another, they are moving so fast that they continue on their way rather than lingering and combining into atoms. Everything changed in 1995 when a team at CERN made the first handful of anti-hydrogen atoms. This was a technological tour de force, which goes beyond the scope of the present book. Suffice to say that everything about anti-hydrogen confirms the perfect overall symmetry between it and hydrogen. The received wisdom is that antiparticles are as likely to be formed as particles, but have little chance of combining to form complex structures before they are annihilated by contact with the all-pervading matter in the environment.

The next simplest antinucleus is anti-deuteron, comprising a single antiproton linked to an antineutron. Individual antineutrons have been produced in experiments, analogous to the creation of antiprotons. But here again, the chance of making both an antiproton and an antineutron, which come into contact so gently that they manage to link into an anti-deuteron before one or other has been destroyed by hitting a nucleus of conventional matter is very small. However, it can happen, and was first observed in 1965.

The most complicated antinuclei made are those of anti-helium-4. Eighteen examples were identified among the debris resulting from the energy released in the collision between two gold nuclei at RHIC in 2011. Had there been a perfect vacuum, the anti-helium would have survived as stably as its familiar material world counterpart, the alpha particle. However, within ten billionths of a second they had crashed into the walls of the experimental apparatus and were destroyed in a miniature fireball.

Everything known about anti-hydrogen, anti-deuteron, and anti-helium is consistent with the theoretical wisdom that the properties of antinuclei match those of their matter counterparts. Thus, if there is antimatter in bulk in remote parts of the cosmos, occasionally one should find evidence of primordial antinuclei in cosmic rays. The search for evidence is a goal of the Alpha Magnetic Spectrometer (AMS) experiment, which is mounted on the International Space Station (ISS).

A purpose of AMS is to detect primordial cosmic rays, before they have hit the atmosphere. All antiparticles that have been seen in cosmic rays are consistent with being produced in violent collisions between the primary cosmic rays, presumed to consist of conventional nuclei and particles, and the atmosphere itself. In such collisions there is very little chance that the basic antiparticles will combine in large clumps to form the antinuclei of heavy elements.

If there are regions of the universe where the stars are made of antimatter, the ensuing supernovae explosions should send the nuclei of anti-elements into the cosmos. Detectors located above the earth's atmosphere should find evidence for any that have survived intact. That is why the AMS experiment is placed on the ISS.

The atomic nuclei of normal matter are positively charged, whereas antinuclei are negatively charged. Thus nuclei and antinuclei will deviate in opposite directions in a magnetic field, which makes it possible to distinguish between them. AMS specifically focused on looking for anti-helium. None has been found; if there is any at all it must occur at less than one part per million relative to helium. To date, all the evidence is that the universe at large is made of matter to the exclusion of antimatter. The periodic table of anti-elements, seeded by nuclei made of antiprotons and antineutrons, exists in theory but there seems no evidence that it occurs in practice in the observable universe.

Antiprotonic helium

Antimatter is thus interesting in its own right, and antiparticles are used to understand details of conventional matter. The insertion of an antiproton into the helium atom, as previously mentioned, and its subsequent annihilation by the nucleus gives information on how complex systems react in the presence of antimatter, and of the preferred pathways by which antimatter annihilates in practice. The process shows how an exotic atom transforms into an exotic nucleus, which then self-destructs. This is an application of antiparticles to the study of atoms and nuclei, which is currently of value to our understanding of fundamental nuclear science.

These examples show how exotic concepts, such as antimatter, have applications in science. We now return to the more familiar world of matter, and conclude this *Very Short Introduction* with a summary of applications of nuclear physics to industry, medical science, and human health.

Chapter 8
Applied nuclear physics

The large amounts of latent energy within the nuclei of
atoms can be liberated in nuclear reactors. Together with
nuclear weapons, this is the most familiar application of
nuclear physics in popular wisdom. The laws of thermodynamics
still apply, which has the consequence that the liberation of
nuclear energy involves change, and the production of ash,
here known as *nuclear waste*. Disposing of this waste, which
can be highly radioactive and a hazard, is a major political
and technological problem. The search for practical power
production by nuclear fusion, using deuterium and tritium, in
hot plasmas or induced by lasers, is an active area of research.
This broad theatre of nuclear power is covered in *Nuclear
Power—A Very Short Introduction*, by Maxwell Irvine. I shall
therefore not include it here, and focus instead on ways in
which the fundamental properties and phenomena of nuclear
physics are used as tools in other areas of science and
technology.

First, let's enumerate the broad opportunities. The phenomenon
of natural radioactivity provides beams of particles, which
may be used to initiate other nuclear reactions, or to attack
tumours in cancer treatment, or to induce radioactivity in a
sample and thereby reveal the presence of elements in minute
quantities.

Radioactive decays occur with characteristic half-lives, and so the measurement of abundances of different isotopes in a sample, when the relative half-lives are taken into account, can give information on the age of the sample. We have already seen how this is used to date rocks, and even the earth; it can be used in forensic science, to distinguish the paint of a modern forger from those of a genuine old master. Nuclear reactions can induce radioactivity in otherwise stable elements. If the ensuing decays involve gamma rays, the spectrum of the energies are like barcodes from which presence of the original element can be deduced, even in minute quantities. This is used in forensic science, and in security systems, to identify the presence of elements that signal drugs or explosives.

Irradiating tumours with gamma rays has been common for decades. However, these damage healthy tissue along the track as well as their goal, the cancerous cells. Protons and other atomic nuclei have electric charge, and so their momentum can be adjusted so that they come to rest at the cancer cells and do maximum damage there. The electric charge of nuclei has another advantage: many nuclei are like miniature magnets, and their magnetism can be used to create images of bodily structures in (nuclear) magnetic resonance imaging (MRI).

Most exotic, perhaps, is that some nuclei are the source of antimatter. Nuclei that emit positrons are used in medical diagnosis as the source for positron emission tomography (PET) scans. Positrons are also used to test materials in industry. So much for generalities; here are some details chosen from a very rich field.

Radiological cancer treatment

Nuclear diagnostic techniques have revolutionized medicine by enabling ways to look inside the body without surgery. Nuclear medicine worldwide is a business with a turnover exceeding

10 billion US dollars. Its importance is primarily based on the radiations emitted by unstable isotopes. These may be used as cures, as in the bombardment of cancer cells by radioactivity from elements that have entered the body, by ingestion or specific implantation, or as beams from an external source. Nuclear radiation is also used as a diagnostic.

The element technetium does not occur naturally on earth, as we saw, because it has no stable isotopes. It is synthesized in nuclear reactions and used widely in medical procedures. Indeed, so important is this element that a strike at a Canadian nuclear laboratory in 1998 threatened over 40,000 medical procedures per day in the USA alone, which led to cancellation of many procedures throughout the world.

A particular isotope, technetium-99m, has a half-life of six hours. (The m denotes this to be an isomer, which is a metastable excited state of the isotope.) It can be injected into a patient, where it becomes a source of gamma rays, which can be detected by a gamma ray camera. Injected into the arteries of the heart, for example, the radioactive decay of technetium enables the blood-flow in coronary arteries to be seen. The radio isotope's half-life is short enough that it is effectively inert after the operation is completed and the patient is not subjected to long-term radiation; its advantage therefore is in enabling data to be gathered rapidly while the total exposure is low. Technetium-99m is widely used to scan bone, the lungs, and the heart.

Forensics via induced radioactivity

One way of using radioactivity to advantage is to induce atomic nuclei already in a sample to become radioactive and thereby reveal their presence and identities. This can be done by the use of neutrons, which can easily penetrate a sample and activate it. The beam of neutrons can be made by using a natural source of alpha particles, which is mixed in with beryllium. The interaction

$$_2^4 He + _4^8 Be \rightarrow _0^1 n + _6^{11} C$$

liberates neutrons, which then irradiate a sample. Slow neutrons are especially effective, being able to enter atomic nuclei without any electrical resistance over long enough timescales for them to be absorbed. This can turn a stable nucleus in some material into a radioactive form, whose subsequent decay can be detected.

Suppose you shine neutrons on a material that contains several elements. The nature of the newly formed radioactive isotopes depends on the original elements. What proves valuable is that the various radioactive species decay at different rates and emit a variety of characteristic radiations. So by monitoring the spectrum of the emissions, we can identify elements even when they are present in minute quantities. The neutrons have activated the original elements, provoking them to advertise themselves. This has applications in forensic science, where traces of foreign elements can be identified and forgeries revealed by the isotopic content of modern materials showing them to differ from the pigments used historically. This has especial application in the art world.

Exposing a painting to a beam of neutrons for about an hour generates a low level of radioactivity within the paints. Any beta rays emitted are recorded. Nuclei of different elements within the paints have different half-lives, so some die out faster than others. Thus nuclei that give up their radiation soonest, with the shortest half-lives, appear only in data taken soon after activation. Longer-lived nuclei, by contrast, show up later, after all others have died away. Different colours of paint involve different elements and show up at characteristic times in the post-activation analysis. The technique can reveal an overpainted signature or even a whole painting that has been obliterated.

Any would-be forger must ensure that the half-lives of paints employed match those of the original.

As neutron sources are compact they can be used in security checks, such as at airports. The idea here is to look for the presence of nitrogen, which is an essential ingredient of plastic explosives such as TNT, nitroglycerine, and nitro-amines (RDX). The neutron activates the nitrogen into an unstable excited state of nitrogen-15

$$^{14}N + n \rightarrow {}^{15}N^* \rightarrow {}^{15}N + \gamma.$$

The gamma ray has a characteristic energy of 10.8 MeV. A gamma ray detector is then programmed to issue an alert when gamma rays of this energy are recorded.

Neutron activation was also used to analyse the hair of Napoleon Bonaparte, who many suspected died from arsenic poisoning. Samples of his hair were used—one from when he was a boy, one during his time in exile, and one at the time of his death. The samples were placed in a nuclear reactor where they were irradiated with neutrons. The only stable isotope of arsenic is arsenic-75. When irradiated with neutrons this converts to arsenic-76, which decays by emission of a gamma ray, with a half-life of about 26 hours. Detection of the gamma rays enabled precise estimates of the amounts of arsenic in the samples to be determined. These showed elevated amounts of arsenic, but these appeared to have been a feature throughout his life, which disfavoured the hypothesis that he had been murdered by exposure to the element.

Diagnostics

We can find out what elements are present on the surface of Mars by inducing them to send us signals in the form of gamma rays. This was the strategy in the Mars Pathfinder mission in 1997, which used a technique known as *Rutherford backscattering spectrometry*. This is also used to diagnose the elemental content of samples in medical biopsies and in industry.

The idea derives from the original experiments of Rutherford's team that discovered the atomic nucleus by its ability to deflect alpha particles. In that original experiment the nuclei were of gold, but suppose instead there had been a variety of elements present. The effect on the alpha particle would depend on the amounts of electric charge and also the masses of the target nuclei. The more massive the target, the more violently the alphas tend to recoil. An americium-241 source of alpha particles was landed on the moon, long before astronauts went there, and alpha particles were bounced off the lunar surface. The numbers were counted and sorted by energy, from which the abundances of elements could be deduced.

In the Mars Pathfinder mission, curium-244 was the source, where it analysed light elements such as carbon and oxygen. Heavier elements, such as sulphur and fluorine, can be identified when the impact of the alpha ejects a proton from the target. The energy of the proton is the key to identifying which element is involved.

Heavier elements, such as iron, can be identified by a third reaction induced by the alpha particles. This is referred to as *pixie*, from the acronym PIXE: particle induced X-ray emission. This involves both atomic and nuclear physics, and arises when the alpha particle (or in laboratory experiments perhaps a proton beam) knocks electrons out of the deeply bound orbits near a heavy nucleus. This leads to the emission of X-rays, whose energies are a direct signature for the elements involved. This technique can be applied to haze in the atmosphere, and can determine the presence of pollutants in levels below one part per trillion. These techniques are also useful in medicine, where the abundance of iron in cells can be determined without destroying the actual cells.

Destruction of healthy tissue can be a problem with chemical techniques, whereas nuclear physics can be a diagnostic with less damage.

Collisions between cosmic rays and atoms in the environment provide the nuclear reactions that can induce radioactivity. We have seen how collisions with nitrogen in the atmosphere leads to carbon-14, whose radioactive decays can identify the abundance of different isotopes of carbon and hence the age of organic matter. This is also exploited in environmental science.

Carbon joins with oxygen to form carbon dioxide, which the atmosphere exchanges with the oceans. These tend to absorb CO_2 near the poles and release it in equatorial regions. The ratio of carbon-14 to carbon-12 falls as time passes, so it is possible to determine how long the carbon-14 has been present in samples at various latitudes and depths, and from this determine the circulation of the oceans and the atmosphere.

Cosmic ray collisions also produce beryllium-7 and beryllium-10. Beryllium isotopes attach readily to aerosols. Aerosols are hosts for the chemical reactions that make forms of chlorine that destroy the ozone. By studying the presence of these isotopes in samples of fallen snow, or of the atmosphere collected by high-flying aircraft, the structure of the ozone layer can be monitored.

Nuclear magnetic resonance imaging

The phenomenon known as nuclear magnetic resonance is used to take images of the human body and diagnose anomalies. It is completely safe, and yet the word 'nuclear' has such a negative impression for many people that it has been dropped from the common description. Today it is known as MRI, magnetic resonance imaging. I find it sad that even nuclear physicists have adopted this politically correct name, and are not robustly advertising the remarkable benefits that nuclear science can give.

Thus I shall assume that readers who have journeyed this far are, like me, prepared to be robust as well as critical and accept my anachronistic use of NMR; if you prefer to use MRI, so be it—the underlying process is NMR, however you choose to name the medical applications.

When electric charges rotate, magnetism can result. Many atomic nuclei have an intrinsic angular momentum, or spin, and being electrically charged produce a magnetic field like a tiny magnet. This property comes into play in NMR.

When a sample of material containing magnetic nuclei is in a magnetic field, the nuclei try to align themselves with the field. As the nuclei are spinning, they don't line up exactly but instead they wobble, or precess, about the field direction, analogous to a spinning top wobbling about the vertical. The rate of this precession depends on the strength of the field and also on the particular nucleus. Thus if you know the field strength, and measure the precession frequency, you can determine the type of nucleus.

To make the nuclei reveal themselves, you apply a small kick, in the form of an electromagnetic wave. When the wave's frequency matches the precession frequency of the nucleus, a resonance phenomenon results. Thus we have all the conditions of NMR: atomic nuclei are magnets and an electromagnetic wave stimulates them to resonance. After this stimulation, the nuclei return to their normal state, which they do by radiating their newly gained energy as radio waves at the precession frequency. Measurement of these frequencies identify the varieties of nuclei.

The application in medicine is based upon the commonest magnetic nucleus in the human body—that of hydrogen, the proton. The technique is to make the protons in a patient's body wobble in a magnetic field whose strength varies across the body. The protons then wobble at different frequencies depending upon their locations. The NMR spectrum thus gives information about the numbers of hydrogen nuclei at different places in the body, which a computer then transforms into an image. This technique can reveal details of soft tissue, which is transparent to X-rays. Furthermore, the amount of radio-wave energy absorbed during an NMR scan are trifling, too small to induce chemical changes in the body.

PET scans: positron emission tomography

When antimatter meets matter, the two can mutually annihilate into a flash of gamma rays. The simplest example is that of an electron being annihilated by its antiparticle, the positively charged positron. If the original electron was quiescent, as in some material, and the positron had little kinetic energy, the result would be a pair of gamma rays, which leave the annihilation site in opposite directions, each with the same energy. Detection of these rays with a special camera can then be used to deduce where the annihilation occurred. In turn, by recording large numbers of such events it is possible to build a map of the distribution of the locations and hence learn about the structure of the original material, such as in a human brain (Figure 13).

This all begs a question: where do you get a source of positrons? Antimatter does not exist in bulk and has to be made, transiently. In the case of positrons this happens in nuclear radioactivity. Beta decay can occur by emitting either an electron or a positron. Recall that if a neutron in one nucleus converts into a proton, electric charge is conserved with the appearance of a negatively charged beta particle—the electron. However, some nuclei become more stable if a proton converts into a neutron, and emits a positron.

Some examples of such 'positron emitters' are carbon-11, nitrogen-13, and oxygen-15. These are radioactive forms of common elements in the human body and can be used, along with positron emission,

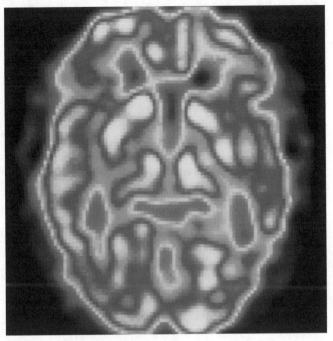

13. PET scan image of a cross-section of a brain.

to trace bodily functions. Radioactive oxygen atoms can be used to label oxygen gas for the study of oxygen metabolism, carbon monoxide for the study of blood volume, or water for the study of blood flow in the brain. If the positron emitter fluorine-18 is attached to a sugar molecule, this can reveal the brain's sugar metabolism and hence the reaction of the brain to various stimuli. Carbon-11, implanted in the chemical, dopamine, is assisting studies of the brain disorders that give rise to Parkinson's disease.

These atoms emit positrons, which are immediately annihilated by electrons in the surrounding tissue. The resulting gamma rays are detected and from this we learn where the source was located.

By surrounding the patient with a halo of cameras, images of the activity can be built up in slices, hence *tomography*—from the Greek tomos, meaning cut (and thus also *atomos*, uncuttable, as the anachronistic source of the word atom).

As these positron emitters often have short half-lives—oxygen-15 is a mere two minutes—they often have to be made at the hospital itself. To do so, a small cyclotron directs beams of protons, deuterons, or alpha particles at suitable targets. Thus, for example, oxygen-15 and nitrogen-13 can be made by the impact of a deuteron on nitrogen-14. The oxygen-15 is made by

$$d(np) + {}^{14}N \rightarrow {}^{16}O^* \rightarrow {}^{15}O + n$$

whereas nitrogen-13 is produced by

$$d(np) + {}^{14}N \rightarrow {}^{16}O^* \rightarrow {}^{15}O + n; \, {}^{13}N + t(nnp)$$

The oxygen-15 can be fed directly from the cyclotron site to the breathing mask of the patient in a nearby room.

While positron emitters are used worldwide in medicine, the technique is also used for some industrial testing of materials. For example, when positrons annihilate with electrons in metals, they can show up metal fatigue. Distortions in the atomic lattice of the metal provide 'resting sites' where the positrons survive slightly longer before annihilating. The observation of this slight delay can identify the onset of metal fatigue before any cracks appear in the material.

These might be said to be the applications of antimatter. However, it is primarily nuclear physics because the positrons only appear courtesy of beta radioactivity, and the short-lived positron emitters are created by nuclear transformation in the first place.

Conclusion

Nuclear physics is a rich and active field. There is much within its vast reach, beyond what is written here: as our title records, this is both an introduction, and very short. We have restricted ourselves to the fundamental ideas, with some mention of history and technique but the experimental apparatus, instrumentation, and techniques would require a much longer text.

We have seen how this field has grown in little more than a century. Atoms were found to have a nuclear centre, which is a new level of reality and the source of vast reserves of energy. The applications of this energy for nuclear power, in industry and in weapons, goes beyond the scope of our emphasis on fundamental ideas. Readers interested in these and other applications of nuclear physics should consult *Nuclear Power—A Very Short Introduction* by Maxwell Irvine and *Nuclear Weapons—A Very Short Introduction* by Joseph Siracusa. The discovery of the atomic nucleus has spawned novel insights and fields of endeavour. The study of nuclear particles, and the use of those particles to investigate the fundamental forces, brings us to the field of particle physics. This outgrowth of early nuclear physics is modern particle physics, and the concepts there are described in *Particle Physics—A Very Short Introduction* by this present author. The applications of radioactivity, and ways to protect against it, are described in *Radioactivity—A Very Short Introduction* by Claudio Tuniz.

We have described the basic ideas of nuclear astrophysics, which is a vast field in its own right, and will be the subject of a separate *Very Short Introduction*. The discovery of the rules that build nuclei led to the empirical mass formula and the consequence that the force of gravity can form huge nuclei of neutrons, which occur as neutron stars. The formation of elements and their relative abundances are now understood thanks to the application of nuclear physics to the cosmology of the big bang and the astrophysics of supernovae. Finally we have the discovery of quarks, with the possibility that there exist forms of nuclear matter much richer than the cold forms that we encounter on earth. That the stable strange matter could be the most stable of all may be science fiction, but the fact that the idea can even be considered is a remarkable testimony to the inspiration of nuclear physics. That the application of nuclear physics in the form of thermonuclear weapons could destroy life on earth is the Faustian bargain that the nuclear genie has released. That we are here at all, however, is thanks to nuclear astrophysics, for as Carl Sagan famously remarked, 'We are stardust'. Or if you are less romantic, we are made from nuclear waste released by a defunct nuclear fusion reactor.

Further reading

The following suggestions for further reading are not intended to form a comprehensive guide to the literature on nuclear physics, but focus on volumes that I have found interesting and informative, and which extend or complement the treatment here. This list includes some 'classics' that are out of print but which should be available through good libraries, or second-hand book sources such as <http://www.abebooks.com>.

James Binney, *Astrophysics, A Very Short Introduction* (Oxford University Press, 2016). An introduction to the physics of stars, in particular the role of nuclear physics.

Brian Cathcart, *The Fly in the Cathedral* (Viking, 2004). The title of my chapter 1 was inspired by this history of how a small group of Cambridge scientists won the race to 'split the atom'.

Frank Close, *Particle Physics, A Very Short Introduction* (Oxford University Press, 2004). The companion to this book, but on particle physics.

Frank Close, *The New Cosmic Onion* (Taylor and Francis, 2007). The basic ideas of nuclear and particle physics in the 20th century, up to the significance of the Higgs boson, for the general reader.

Frank Close, Michael Marten, and Christine Sutton, *The Particle Odyssey* (Oxford University Press, 2003). A highly illustrated popular journey through nuclear and particle physics of the 20th century, with pictures of nuclear and particle trails, of experiments, and of people.

Maxwell Irvine, *Nuclear Power, A Very Short Introduction* (Oxford University Press, 2011). An introduction to some applications of nuclear physics, which is complementary to the present book.

John Polkinghorne, *Quantum Theory, A Very Short Introduction* (Oxford University Press, 2002). An introduction to quantum theory, which describes the behaviour of the subatomic world and the atomic nucleus.

Richard Rhodes, *The Making of the Atomic Bomb* (Simon and Schuster, 1986). The authoritative story of the Manhattan Project, with a superb history of the early days of nuclear physics.

Claudio Tuniz, *Radioactivity, A Very Short Introduction* (Oxford University Press, 2012). An introduction to the basic ideas of radioactivity, together with its implications and applications.

Steven Weinberg, *The First Three Minutes* (Pantheon Books, 1992). After more than two decades, this remains the outstanding popular exposition of creation of the elements in the aftermath of the big bang.

W.S.C. 'Bill' Williams, *Nuclear and Particle Physics*, revised edition (Oxford University Press, 1994). A detailed first introduction suitable for undergraduates studying physics.

David Wilson, *Rutherford—Simple Genius* (Hodder and Stoughton, 1983). A thorough biography of the father of nuclear physics, with chronological history of the early decades of the field.

Index